GROWING
UNDER
COVER

GROWING
UNDER
COVER

Techniques for a More Productive, Weather-Resistant,
Pest-Free Vegetable Garden

NIKI JABBOUR

Storey Publishing

The mission of Storey Publishing is to serve our customers by publishing practical information that encourages personal independence in harmony with the environment.

Edited by Carleen Madigan
Art direction by Carolyn Eckert
Book design by Carolyn Eckert and Erin Dawson
Indexed by Christine R. Lindemer, Boston Road Communications

Cover photography by © Jeff Cooke/Cooked Photography
Interior photography by © Jeff Cooke/Cooked Photography and © Niki Jabbour, 11 l., 15, 19 t., 29, 42, 48, 52, 58, 61–65, 67, 68, 74, 75, 80, 84, 94 t., 101 l., 102, 118 r., 129, 132 r., 134 t., 135, 137 t., 142 l., 147 t., 148, 151, 154, 158, 160 r., 164 t., 165 t., 167 r., 173, 176 r., 182 r., 192 l., 197, 201, 205; © Niki Jabbour, taken at Watershed Farm, 60, 123, 165 b., 182 l., 188 l.
Additional photography by © Agriology/Alamy Stock Photo, 133; © Avalon/Photoshot License/Alamy Stock Photo, 107 b.l.; © Besjunior/Alamy Stock Photo, 107 b.r.; © Brenda Franklin, 44, 89; © Clearskiesahead/ iStock.com, 55 l.; © GordonImages/iStock.com, 126; © Hartley Botanic, 46, 54; © inomasa/iStock.com, 107 t.l.; © Island Images/Alamy Stock Photo, 92; © Janet Horton/Alamy Stock Photo, 105 t.l.; © jess311/iStock .com, 134 b., 137 b.; © Jessica Walliser, 121 r.; © John Glover/GAP Photos, 161; © Joseph De Sciose, 28, 37 l., 39, 41; © Kay Roxby/Alamy Stock Photo, 132 l.; © Ken Leslie/Alamy Stock Photo, 186; © Loop Images Ltd/ Alamy Stock Photo, 111; © Nan Sterman, 2019, 18; © Pavel Abramov/iStock.com, 136 b.r.; © Regenerative Design Group, 70, 71; © sasimoto/stock.adobe.com, 170; © schankz/Shutterstock.com, 134 m.;© Sergey_ Fedoskin/iStock.com, 107 t.r.; © Slavomira Kovacova/ iStock.com, 181 m.; © slertwit/123RF.com, 131; Courtesy of Steve Farley, The Optimistic Gardener, 56; © Tara Nolan, 37 r.; © Veg Organic/Alamy Stock Photo, 147 m.; © Westend61 GmbH/Alamy Stock Photo, 171; © World History Archive/Alamy Stock Photo, 21; © y-studio/iStock.com, 55 r.

Storey Publishing
210 MASS MoCA Way
North Adams, MA 01247
storey.com

Printed in China through World Print
10 9 8 7 6 5 4 3 2 1

Storey books are available at special discounts when purchased in bulk for premiums and sales promotions as well as for fund-raising or educational use. Special editions or book excerpts can also be created to specification. For details, please call 800-827-8673, or send an email to sales@storey.com.

Library of Congress Cataloging-in-Publication Data

Names: Jabbour, Niki, author.
Title: Growing under cover / Niki Jabbour.
Description: North Adams, MA : Storey Publishing, [2020] | Includes index.
Identifiers: LCCN 2020028928 (print) | LCCN 2020028929 (ebook) | ISBN 9781635861310 (paperback) | ISBN 9781635861327 (ebook)
Subjects: LCSH: Vegetable gardening. | Greenhouse gardening. | Cloche gardening. | Cold-frames.
Classification: LCC SB321 .J3195 2020 (print) | LCC SB321 (ebook) | DDC 635—dc23
LC record available at https://lccn.loc.gov/2020028928
LC ebook record available at https://lccn.loc.gov/2020028929

For Dany, Alex, and Isabelle

And for all the gardeners who deal with frost, hail, wind, drought, cold, heat, insect pests, deer, groundhogs, and squirrels. This one's for you.

CONTENTS

PART 2: Vegetables That Love a Cover 140

Why I'm an
UNDER COVER
GARDENER

I am an under cover gardener. And no, that doesn't mean I'm a garden spy covertly peeking through the hedge at my neighbor's veggie patch. It means I'm a gardener who uses simple but effective covers to grow more food.

My own introduction to gardening under cover began with a pop-up polytunnel. I was only 16 years old but was already starting seeds indoors, growing herbs in pots, and had taken control of the family vegetable garden. The simple structure, which was a gift from my parents, was really more of a small, clear tent, measuring just 6 by 8 feet. And boy, did I love it.

That first spring, I stuffed it full of all the seedlings I had grown on my mother's dining room table (hey, now I get why they gave me a polytunnel!). It didn't take long for me to learn that a covered structure warmed up quickly, even on cold days, and produced sturdy, stocky seedlings.

I had a lot of fun with that little polytunnel, but it was flimsy and only lasted a few years. Soon, I was off to university and it would be 7 years before I had my own garden space again. However, once I was back in the garden, I quickly rediscovered just how handy season extending garden covers could be.

In my first book, *The Year-Round Vegetable Gardener,* I shared some of my favorite techniques for harvesting homegrown vegetables and herbs in all seasons (including winter!). Today, I still use those methods, but I've also added new types of covers and structures to my food garden, to boost yields and grow healthier, higher-quality vegetables. I use more shade cloth in my summer garden to extend the harvest of cool season greens. I switched from PVC to metal hoops for my mini hoop tunnels because they're much stronger, and I added a 14-by-24-foot unheated polytunnel.

The tunnel offers plenty of space for year-round harvesting, but it's also become my garden sanctuary; I can escape to the veggie patch no matter the outside weather. It's almost always spring in the polytunnel. And with that in mind, we gave up a bit of growing space in the back corner of the tunnel to create a small indoor patio. It's the perfect spot to sow seeds, get some writing done, or just enjoy a cup of tea surrounded by a jungle of veggies.

That said, you don't need a big space or a huge garden to use the methods featured in this book. Many of my under cover techniques can be used in super small gardens, a single raised bed, or even containers on a deck or patio. You also don't need to spend a lot of money to become an under cover gardener. There are many inexpensive — or free! — types of covers that you can put to work in your garden.

I've learned a lot about matching covers to crops. The types of covers and structures I use depend on the season and the crops being grown. Of course, there's also a lot of overlap, as many of the covers can be used in more than one season. For instance, I use row covers in spring, summer, fall, and winter, but shade cloth is really just for summer.

Gardening under cover is not just about sheltering crops from cold or heat. It can also be a temporary fix for inclement weather like hail, downpours, or strong winds. Or a way to keep birds, deer, groundhogs, rabbits, and other pests away from your vegetables. I battle deer on a daily basis, and a simple length of netting floated on mini hoops helps give young seedlings a deer-free head start.

Covers also help keep insects like flea beetles, cabbage worms, and cucumber beetles from decimating my crops. They can even reduce the occurrence of diseases like tomato blight, which is devastating to garden-grown tomatoes; those grown in greenhouses or polytunnels are less susceptible to this common fungus.

If you're new to garden covers, this may all sound like a lot of work, but trust me — it's not. I'm not one to fuss over my plants, but using covers to extend and protect the harvest is both easy and effective. It's turned our traditional summer vegetable garden into a year-round food factory!

The garden covers featured in this book are the perfect complement to a home vegetable garden. If you're new to food gardening, there's no need to jump right into a greenhouse or polytunnel. Start with a row cover or mini hoop tunnel, graduating to a cold frame after a year or two. As you gain skills, confidence, and experience, you may discover that you're ready to tackle a small backyard dome or a DIY polytunnel.

What should you grow within these covers? Grow what you like to eat! But should you want more specific information, I've shared the growing details of a wide range of vegetables in part 2, and listed all the garden covers I use for each vegetable.

In the end, my message is simple: garden covers are the easiest way to grow more food for a longer period of time, grow higher-quality plants bothered by fewer pests and diseases, and mitigate the stresses caused by cold, heat, and extreme weather.

EIGHT REASONS
to Become an Under Cover Gardener

GROW MORE FOOD. We all want to be as efficient and grow as much food as possible in our gardens; using covers and covered structures will allow you to do just that.

CONTROL THE ENVIRONMENT. Depending on the crop, your vegetables can be susceptible to heat, cold, wind, hail, snow, and other environmental conditions. Using covers or structures to create microclimates aids in preventing damage and maintaining crop quality.

HARVEST YEAR-ROUND. Being able to provide homegrown food for my family 365 days a year is something that I'm proud to brag about. But the truth is that it's not that hard. Covers allow me to harvest sooner in spring, later in fall, and throughout the winter.

SAVE MONEY. Like most gardeners, I'm budget minded and don't like wasting money — or food! Because growing under cover helps me increase yields and enjoy a year-round harvest, I'm shaving serious dollars off my weekly grocery budget.

REAP A HYPERLOCAL HARVEST. Moving food around the world creates an incredible amount of greenhouse gas emissions. If I go to my supermarket any month of the year, there's a good chance that

the plastic tub of lettuce has been shipped to my Nova Scotia grocery store from California. By growing it myself, I'm cutting the distance our food has travelled from thousands of miles to mere feet.

REDUCE PEST PROBLEMS. I'm not going to say you won't have any pest problems (let's be realistic), but you will have fewer pest problems when you use certain covers. Insect barriers, row covers, deer fencing, deep mulching, cold frames, and polytunnels are simple covers that can prevent insect and animal pests from eating your vegetables.

GROW THE WORLD. Using season-extending covers like mini hoop tunnels and polytunnels has allowed me to grow a wide variety of global vegetables like edible gourds and cucumber melons that are normally difficult to grow in a short-season region.

DIY YOUR WAY TO MORE FOOD. Some of the more serious structures, like geodesic domes, polytunnels, and greenhouses significantly increase the amount of food you can grow, but they can be expensive to buy. However, handy gardeners will find that they can DIY many structures using commonly found materials like lumber, PVC conduit, polycarbonate, and polyethylene.

PART 1 \ Introducing the Covered Garden

Garden covers such as row covers, mini hoop tunnels, and cold frames have transformed my vegetable garden. They shelter crops from weather extremes, helping mitigate the new reality of a changing climate, as well as protect against insects, deer, rabbits, and birds. By creating microclimates in the garden, they also extend the season on both ends of the summer and help create better conditions for seed germination and for new seedlings to take root. There's not a day of the year where I'm not using some type of cover to grow better crops, whether it's the new seedlings sheltering under simple cloches, tomatoes basking in the warmth of the polytunnel, arugula thriving without flea beetles with the protection an insect barrier, or hardy winter greens awaiting harvest in the dead of winter.

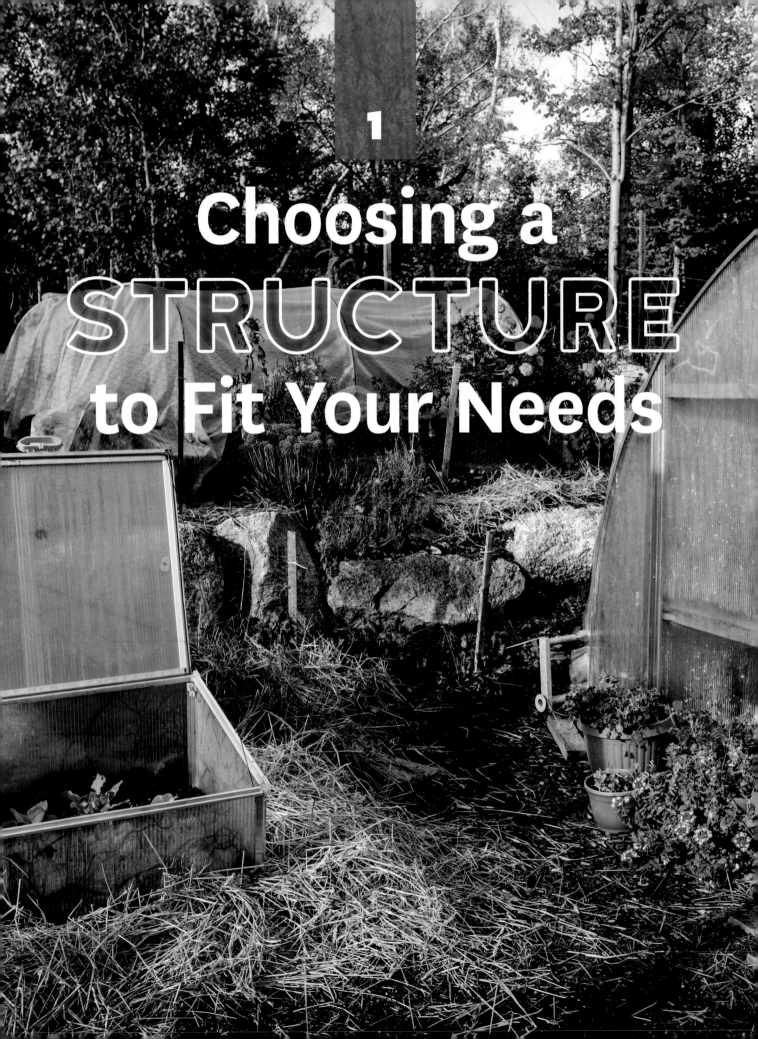

1

Choosing a
STRUCTURE
to Fit Your Needs

arden covers aren't just for growing into winter or protecting from frost. They can be used all year long to boost yields, protect from pests, reduce disease, establish summer sowings, and improve crop quality. I use a variety of fabrics, devices, and structures to capture heat, reduce heat, provide shade, and create a barrier against pests. Most of them, like shade cloths, row covers, and mini tunnels, are inexpensive or easily made. A few, like polytunnels and greenhouses, require a larger investment.

POLYTUNNEL

MINI HOOP TUNNEL

CLOCHES

WHAT TYPE OF PROTECTION DOES YOUR GARDEN NEED?

What type of structure you ultimately choose depends on what your goals are. Do you want an extra-early harvest of spring greens? Do you live in a short-season region where it's difficult to mature heat-loving vegetables like tomatoes and peppers? Or are your summers hot and dry, making it hard to establish crops for fall and winter harvesting? Maybe your garden is plagued by deer, rabbits, cabbage worms, or other garden pests. Before you invest in a cover, consider why you need it.

Protection from Weather Extremes

A protective cover is any material or device that shelters plants from weather — cold, frost, wind, snow, hard rain, and hail. Structures like cold frames or mini hoop tunnels collect and trap solar energy, creating a microclimate around the vegetables inside. They can extend the harvest by weeks or months in autumn or push up the planting season by the same time frame in early spring. They can also provide warmer growing conditions, allowing gardeners in colder regions to cultivate plants like eggplants and melons.

BEST OPTIONS

- ▶ Cloches for temporary cover
- ▶ Mini hoop tunnel with row cover or plastic
- ▶ Polytunnels
- ▶ Greenhouses and geodesic domes

In winter, mini hoop tunnels cover my raised beds and a cold frame is mulched with evergreen branches for added insulation. Inside the polytunnel, a row cover helps insulate winter salad greens.

Garden designer Nan Sterman protects her own vegetable garden from critters with a hoop house covered in half-inch-mesh hardware cloth. The mesh allows light, air, and rain to enter, but keeps out larger garden pests.

Defense against Pests

In my garden, thin insect barrier fabrics keep pests such as cabbage worms, Colorado potato beetles, squash bugs, and cucumber beetles from eating my crops. Combining fabrics with crop rotation is a smart way for organic gardeners to thwart these common pests, but the fabrics need to be applied and removed at the right times. Apply them in early spring as soon as the seeds or seedlings have been planted. If you wait to lay your fabrics, you risk having the newly emerged adult insects find their target plants before you cover them. In that case, you're just creating a predator-free environment for them to nibble on your veggies.

CHOOSING A GARDEN COVER

	Insect barrier	Row cover	Shade cloth	Cloche	Mini hoop tunnel with polyethylene cover	Cold frame	Polytunnel/ Greenhouse/ Dome
FROST PROTECTION	N	Y	N	Y	Y	Y	Y
SPRING/FALL MICROCLIMATE	Y	Y	N	Y	Y	Y	Y
SUN PROTECTION	Y	Y	Y	N	N	N	N
PEST PREVENTION	Y	Y	N	N	Y	N	N
DISEASE PREVENTION	N	N	N	N	N	N	Y
WINTER HARVESTING	N	N	N	N	Y	Y	Y

Floating fabric on hoops above, rather than directly on top of, the plants looks tidier and prevents bunching, which can block light. Once the fabrics are on the beds, they can be left in place until the plants begin to flower (in the case of squash and cucumbers) or until harvest (in the case of broccoli, kale, or potatoes). Keep in mind that it's also important to rotate crops each season to avoid a repeat of insect infestations, diseases, or nutrient depletion.

Lightweight fabrics also make a reliable defense against larger pests like deer, rabbits, and birds. I often drape insect barriers or bird netting over my newly seeded beans, peas, corn, and lettuce, which seem particularly tempting to the local bird population. The covers are left in place until germination. They can even speed up the germination process because they help retain heat and moisture.

BEST OPTIONS
► Mini hoop tunnel with row cover, insect barrier fabric, or bird netting
► For deer, rabbits, and birds: insect barrier fabric, row cover, bird netting, chicken wire
► For insects: insect barrier fabric

Shade from the Sun

For most crops, full sun (at least 8 hours per day) is necessary for healthy growth and high production. But sometimes intense sun and heat is too much for certain crops, like cool-season salad greens. If you live in a warm climate, shade cloth is a useful tool for growing a variety of vegetables. Even in my northern garden, I use it in late spring and summer to slow the bolting of my favorite salad crops like spinach, arugula, pak choi, and lettuce.

There are different weaves of shade cloth, which block different percentages of light. Common percentages include 30 and 50 percent. Shade cloth is generally floated over a garden bed to prevent heat buildup. I float shade cloth on my hoops, because it's a quick and easy way to hang the cloth, but it can also be attached to stakes at the corners of a bed. Lengths of shade cloth often come with grommets for easy hanging, but clip-on grommets are also easily available.

BEST OPTIONS
► Temporary shade from an upturned laundry basket
► Mini hoop tunnel with shade cloth
► A shade house or a lath house
► A row cover

Many garden supply companies offer mini tunnels with interchangeable mesh and plastic covers for excluding pests or protecting from inclement weather.

Willow cloches or simple upturned baskets offer needed shade to newly planted seedlings.

USING STRUCTURES STRATEGICALLY

Layering Structures

Layering one protective structure inside another — a mini hoop tunnel layered inside a polytunnel, for example — offers additional warmth. Garden fabrics like row covers, insect barrier, and shade cloth are perhaps the easiest covers to layer inside polytunnels, greenhouses, domes, mini hoop tunnels, and cold frames. They're inexpensive, quick to install, and easy to remove.

Layering covers can be temporary and left on for just a few hours, days, or weeks, or for an entire season. I find layering particularly useful in early spring when the temperature inside my polytunnel can vary from 18°F (–8°C) in the morning to 95°F (35°C) by mid-afternoon. I open the structure in midmorning to vent as the temperature begins to climb, but by late afternoon the temperature starts falling and it's time to roll down the sides or close up the windows and add a layer of row cover on top of young salad greens and root crops. I also layer fabric-covered mini hoop tunnels in the winter polytunnel over salad greens, root crops, and hardy herbs and include a layer of straw or row cover in a winter cold frame for added insulation over root crops.

Setting Up Temporary Structures

Sometimes covers are applied in sequence to a garden bed. For example, in early spring when more protection is needed, polyethylene covers my mini hoop tunnels. As the weather warms, I replace the plastic with a medium- or lightweight row cover. If I have insect-prone vegetables like cabbage or broccoli in the bed, that row cover can be replaced with an insect barrier fabric. If the crop in the bed is a cool-season salad green like arugula, lettuce, or spinach, I may switch out the row cover for shade cloth to delay bolting and extend the harvest as the weather warms.

Gardening in raised beds makes erecting temporary season extenders a snap. My beds are 4 feet wide which is an ideal size for mini hoop tunnels (whether covered with fabric, bird netting, polyethylene, or another material).

Keeping your materials organized in a handy location like a shed or garage also saves time. Fabric and plastics should be cleaned, folded, and stacked between uses, and materials for making the hoops, wire arches, PVC conduit, or metal conduit should be taped together and set aside neatly. If inclement weather is on the way — wind, hail, snow, rain, for example — it takes me very little time to gather my materials and set up my structures.

On their own, garden structures are versatile garden tools. But they can also be combined in strategic ways to increase cold weather protection or to prevent pest damage.

A Short History of
GROWING UNDER COVER

Protected gardening is not a new idea; gardeners and farmers have been producing food under different types of covers for millennia. The first reference to using a greenhouse-like structure for growing food comes from Pliny the Elder in the first century CE. He describes structures called "specularia" that were built to grow cucumbers or cucumber-like fruits year-round for Emperor Tiberius.

Specularia were basically garden beds covered with frames coated in transparent mica, which both captured heat and allowed light to reach the plants. The emperor's specularia were mounted on wheels so that they could be moved into the sun each day and brought back indoors each night.

Fast-forward to mid-fifteenth-century Korea, where simple temporary structures were used to grow vegetables and fruits in winter for the royal household. Described in the Annals of the Joseon Dynasty, these structures were erected over fruit trees for winter protection, but also over vegetables produced during the winter. The structures were built on a sunny site with south-facing windows to capture the sun. On warm days, the windows were left open, and when the weather was cold, straw mats were placed over the closed windows to retain heat.

Even the cloche has been an important part of food gardening for centuries. Cloches first emerged in Italy during the early seventeenth century and made it possible for farmers and growers to start crops earlier in the season. In the eighteenth and nineteenth centuries, European kitchen gardeners harvested crops out of season by using manure-heated hotbeds, glass cloches, and small greenhouses covered in oiled materials like fabric and heavy paper. English gardeners used glass-covered cucumber frames to get an early start on the growing season.

Garden covers eventually made their way across the Atlantic Ocean to Mount Vernon and Monticello, where George Washington and Thomas Jefferson used hotbeds and greenhouses to grow vegetables, citrus fruits, and other plants out of season.

The advent of plastic was a game changer for garden covers. It led to improved materials, larger structures and season extenders that were affordable and effective. Row covers, PVC, polyethylene, and polycarbonates have made it possible for me to harvest a wide variety of vegetables and herbs year-round from my northern garden.

2

STARTING SMALL

Mini Hoop Tunnels and Cold Frames

There is no one right garden structure. If you're new to gardening, start small and begin with a mini hoop tunnel or a cold frame. This will give you an opportunity to flex your gardening skills and learn techniques like timing off-season growing and how to regulate temperature by venting regularly.

Mini hoop tunnels and cold frames also require little initial investment of either time or money. They can be built using new materials or from old windows, doors, bricks, conduit, and even straw bales. For me, the biggest advantage of these devices is that they are fairly easy to build — especially the mini tunnels — and you see their effectiveness almost immediately. Within a week of sowing seeds in my spring frames, tiny seedlings are popping up. I get to enjoy a full crop of salad greens before my neighbors have even begun to prep their gardens.

Once you've had a season or two under your belt, you'll know if you're ready to move to a bigger structure like a polytunnel or greenhouse.

MINI HOOP TUNNELS

A mini hoop tunnel is exactly what it sounds like: a miniature polytunnel. But unlike a walk-in structure, these pint-size tunnels are quick to build, easy to use, and made with inexpensive materials. There are just two main components: hoops and a cover. The hoops can be made from a variety of materials, including PVC conduit, metal, wire, concrete reinforcement mesh, or even old Hula-Hoops cut in half.

Five Ways to Use a Mini Hoop Tunnel

Frost protection. Top your hoops with a row cover to protect from frost or cold weather. A row cover also shelters crops from heavy rain, hail, or strong winds.

Winter harvesting. We use mini hoop tunnels to harvest cold-season crops all winter long. For winter crops, cover the hoops with a 6 mil

A temporary mini hoop tunnel made from nine-gauge wire and row cover fabric will protect crops during the main growing season. Those made from sturdier materials like metal conduit are great for long-term winter protection of cold-season crops.

greenhouse polyethylene. Twist the ends closed and weigh down the sides to secure them against the winter weather.

Summer shade. I use my mini hoop tunnels from late spring to early autumn to provide some shade from the hot sun. Cool- and cold-season vegetables like salad greens quickly bolt once spring turns to summer. Having a length of shade cloth over the hoops lowers temperatures and prolongs the harvest season. You can also use shade cloth to establish just-planted seeds or seedlings.

Spring, summer, and autumn insect defense. Using a lightweight insect barrier keeps cabbage, kale, broccoli, potatoes, and other pest-prone plants free of insect damage. Float the cover over the hoops as soon as crops are planted and bury the sides to prevent pests from entering the mini tunnel.

Spring, summer, and autumn pest prevention. Not all pests are small. Deer, rabbits, birds, chickens, and even dogs can eat or damage vegetables. Top the mini hoops with bird netting or chicken wire to keep crops safe.

Types of Mini Hoop Tunnels

In my mind, mini hoop tunnels fall into two categories: lightweight and heavy duty. Your reason for covering the crop and the timing of coverage help determine the type of hoop you'll need. Are you prewarming spring soil, shading summer-planted seeds from sun, protecting plants from fall frost, or creating a winter harvest tunnel?

Lightweight tunnels. I make lightweight tunnels with wire hoops and use them as spring and fall frost protection in the garden. They do a fine job of protecting crops from cool weather, light frost, heavy rain, and wind, but don't stand up to snow. I also place them over the beds inside my polytunnel for a double layer of winter protection. For this light work, wire or PVC hoops are fine.

I generally prefer to build my own mini hoop tunnels, as it takes little time and I can match the size to my raised beds, but there are many commercially produced mini hoop tunnel kits you can buy. The kits often have wire hoops and fall into the "lightweight tunnel" category. Depending on

HOW HIGH
Should Your Tunnels Be?

Just-planted seeds or seedlings, or compact crops like leaf lettuce, mâche, or baby greens don't require a high hoop. But taller edibles like kale, collards, leeks, and Italian parsley can grow several feet high, so take plant size into consideration when deciding on how big a hoop to use.

In general, hoops should be tall enough so that plants aren't in direct contact with the cover. The exception to this is when lightweight insect barrier fabric is laid on vegetables during the growing season for pest control. Otherwise, plants shouldn't come in contact with a polyethylene cover or row cover during cold weather as this can cause damage to the foliage. This also holds true in summer when a shade cloth-topped mini hoop tunnel is used to shield vegetables from the hot sun. Keep the cover well above crops to avoid heat buildup around the plants.

the manufacturer, mini hoop tunnel kits may be called polytunnel cloches, mini greenhouses, or mini tunnels. Often the kits are too narrow for my beds, or not long enough.

Heavy-duty tunnels. The other type of mini hoop tunnel is a heavy-duty version made from 10-foot lengths of ½-inch diameter PVC or metal conduit. These are strong enough to withstand a snow load and are great for winter protection; if you live in an area that gets a lot of snow, consider adding an extra center support.

Installing Hoops

Nine-gauge wire. For quick hoops, use 9-gauge wire, which is readily available from a hardware store. You can buy the wire in 50-foot coils and cut it to the desired length with wire cutters. The length of the wire will depend on how wide the bed is and how high you want the hoop to be.

For my 4-foot-wide beds, I cut the wire into 8-foot lengths. This creates hoops that are 3 feet tall before the ends are pushed down into the soil. After that, the hoops are about 2½ feet tall.

For my 3-foot-wide polytunnel beds, I cut the wire into 6-foot lengths and the hoops are 2 feet tall at the center before being inserted into the soil. Once the hoops are installed, they're about 15 to 18 inches tall. These low hoop tunnels are used for sheltering low-growing greens like arugula, leaf lettuce, and Asian greens.

Flexible 9-gauge wire can also be bent by hand into a square-shaped hoop, so that the tunnel is the same height over the entire bed.

PVC conduit. For over a decade, ½-inch-diameter PVC conduit has been the primary material I've used for my mini hoop tunnels. Ten-foot lengths are readily available, inexpensive, easy to use, and durable.

To install hoops, pound foot-long rebar stakes into the ground every 3 to 4 feet along each side of the bed, then bend the length of PVC over the bed and slip the ends over the rebar stakes to secure the hoop to the ground. For winter tunnels in areas like mine that have a heavy snow load, a PVC structure needs a center support to add strength. Without it, the tunnels will be susceptible to flattening after a very heavy snow.

Metal conduit. Half-inch-diameter metal conduit makes for a very sturdy hoop to support fabric- or plastic-covered mini hoop tunnels. Metal hoops can be used year after year and are strong enough to withstand snow loads without the center support needed for PVC hoops. To further improve their capacity, I make sure the cover is pulled taut and well secured at the ends, and I knock off heavy snow after a storm. Ten-foot lengths of metal conduit are easy to source at most building supply or hardware stores, and they don't require a stake to hold them in place; just sink the end of each hoop 6 to 8 inches into the soil, spacing the hoops 3 to 4 feet apart.

The only challenge is figuring out how to bend the metal conduit into hoops! A few years ago, I got a metal bender from Johnny's Selected Seeds and started bending 10-foot lengths of conduit into super sturdy hoops. This tool has been a game changer. Plus, bending the metal conduit to make the hoops is great fun. It took just a few minutes to get the hang of it and now I can bend a hoop in about a minute.

To make a metal hoop, you first mount the bender to a sturdy structure like a workbench or a picnic table. Then insert a straight length of conduit into the bender and pull the metal tubing down toward you, bending it into the characteristic U-shape.

A metal bender is a great investment for a garden club, urban farm, or community garden where many gardeners can use it to bend strong hoops for season extension.

PVC CONDUIT

BEND YOUR OWN MINI HOOPS

My low tunnel hoop bender makes quick work of bending metal conduit for 4-foot wide hoops. The bender needs to be mounted on a sturdy surface like a truck hitch or, in my case, a heavy pallet anchored to a wood base.

1 For a 4-foot wide hoop, slide a 10-foot length of half-inch or three-quarter-inch metal EMT conduit through the holding strap, so it extends 16 inches beyond the end of the bender.

2 Grab the opposite end of the conduit and bend it toward you until it touches the base of the bender.

3 Now slide the tube through the holding straps until it's centered on the bender.

4 Squeeze each end of the tube until they are straight and a U-shape has been achieved. The sides of the tube should be parallel. Once you get going, it will take around a minute to bend each hoop.

CONCRETE MESH TUNNELS

A 4-by-8-foot sheet of concrete reinforcing mesh is another material that can be used to make hoops for a mini hoop tunnel. Bend the panel over the bed to provide support for a lightweight cover like row cover, polyethylene, or shade cloth. It is easiest to install mesh tunnels in raised beds that have sturdy wooden edges. I secure the mesh with an untreated 1×2 cut to length and screwed to the bed. Larger mesh cattle panels are often used to DIY a small polytunnel.

Choosing Covers

As with the hoops, there are different options for the cover of a mini hoop tunnel, and I base my mini hoop tunnel cover choice on the season and the use. It makes a difference if I'm trying to protect crops from frost or sun, prewarming the soil, or giving heat-loving vegetables like eggplants or melons a bit of extra warmth.

PLASTIC

I learned the lesson "You get what you pay for" a long time ago when I used inexpensive construction-grade plastic for my winter mini hoop tunnels. The thin plastic couldn't stand up to the winter weather and the ripped covers were soon flapping in the wind. Today I only use greenhouse polyethylene. It does cost more, but with proper care it can last for years. Thinner plastics are best reserved for light protection in spring, summer, or early autumn.

Construction-grade plastic sheeting. These thin, non-UV-treated films aren't long-term covers and degrade after just a few months. But they are inexpensive and can be used for short-term spring or fall frost protection, prewarming the soil, or creating a microclimate over heat-loving crops.

Preslit or perforated polyethylene. I've only tried a few of these products, which come with small holes or slits in the polyethylene to self-vent and regulate temperature. The films are generally quite thin, just 0.8 mil thick, and are useful for planting crops about a month earlier than normal planting dates.

Greenhouse polyethylene. Many garden centers or supply stores offer 4-year, 6 mil, UV-treated polyethylene by the running foot or by the roll. If you're just making a few mini hoop tunnels, buying by the running foot is likely the cheapest way to go. However, if you've got a bunch of tunnels to build or are part of a community garden or urban farm, you might consider buying a roll and sharing the cost. The cost per square foot is much cheaper when you buy it by the roll. For more on greenhouse polyethylene, turn to page 56.

ROW COVERS

These are lightweight fabrics made with spunbonded translucent polypropylene, which allows light, air, and water to pass through to the plants. Row covers have been one of the most essential tools in my garden for the last two decades. They are super effective at protecting crops from frost and cold weather, as well as from insects, birds, deer, rabbits, and other garden pests. You can also use row covers to hold insulating winter mulch in place over root vegetables or to isolate crops for seed saving.

There are four main weights of row covers. The lighter covers allow more light to pass but retain less heat, while the heavy-duty covers have low light transmission but hold more heat. The one you choose will depend on how you plan to use it. Are you sheltering salad greens from a potential frost? Trying to prevent cabbage moths from laying eggs on your broccoli plants or Colorado potato beetles from decimating your potato crop? Or do you wish to shield vegetables from the hot summer sun? In general, I find lightweight and medium-weight row covers are the ones I use the most; I save the heavier covers for winter, when plants aren't actively growing.

Insect barrier. This is the lightest-weight row cover (0.45 ounces per square yard) and is an effective barrier against common garden pests like Colorado potato beetles, flea beetles, and imported cabbage worm, not to mention larger pests like birds, deer, and rabbits. It also prevents the transmission of diseases some insects carry (I'm looking at you, cucumber beetle and flea beetle!). Preventing or reducing pest populations with an insect barrier fabric can also mean decreased populations in subsequent years.

Insect barrier allows around 90 percent light transmission, and air and water can easily pass through the thin fabric. Because it offers good air exchange, heat doesn't build up easily under the cover. It's the best fabric for preventing summer insects from eating heat-loving crops like tomatoes, peppers, and eggplant. However, this flimsy fabric offers little in the way of frost protection. If it's all you have on hand with a frost in the forecast, double it up to boost its frost-busting capability.

Insect NETTING

Insect screens, meshes, and nettings are lightweight covers designed to allow light, air, and water to pass through but prevent a wide variety of insect pests from damaging your vegetables. They also keep out larger pests like deer, rabbits, groundhogs, birds, and squirrels, or pets like dogs and cats. To be effective as a pest barrier, the bottom of the netting must be buried, weighed down, or otherwise secured. They can be left in place temporarily or for the entire season; just be sure to allow enough slack to accommodate growing plants.

Netting is more transparent than row cover fabric, and therefore blocks less light. It can be laid directly over plants, but I prefer to float it over beds on wire or PVC hoops, cages, or other types of support. It looks tidier and offers better protection from larger pests.

Insect barrier is also used by seed savers who need to isolate crops and prevent cross-pollination in crops like cucumbers, melons, and squash. But if you're protecting crops that need to be pollinated in order to produce their crop, be sure to remove the covers when the plants begin to flower.

Lightweight. Lightweight row covers (0.55 ounces per square yard) are what I use most in my garden. They're thin enough to let light pass through (around 85 percent), but thick enough to protect crops from frost. In fact, they're effective down to 28°F (−2°C) and can be left in place for weeks in the spring and fall garden. When the daytime temperature is mild, remove them to encourage good air circulation.

These sturdy covers are also useful for sheltering young seedlings from inclement weather: hail, heavy rain, and strong winds. Unlike insect barrier fabrics, these covers provide some heat retention. In a pinch, you can use lightweight row cover as an effective pest control, but because it isn't as translucent as insect barrier fabric, it may slow the growth of your plants.

Medium weight. Not as versatile as a lighter material, medium-weight row cover (0.9 ounces per square yard) is still useful to have on hand as a temporary frost protector. It insulates vegetables to 26°F (−3°C), but the heavier weight only allows 70 percent of light to pass through. As such, I don't recommend using medium-weight covers as long-term protection during the growing season. However, you can use them as a short-term safeguard in spring and summer over freshly seeded beds, just-planted seedlings, or established crops to prevent damage from deer, rabbits, birds, frost, or bad weather.

Once late autumn arrives and the day length shrinks below 10 hours (early November in my region), I float medium-weight row covers over autumn crops or beds of overwintering greens. Medium-weight covers resist tearing better than insect barrier and lightweight fabrics, making them a sturdy cover for laying on top of deeply mulched beds of winter root and stem crops like carrots and leeks.

Heavy weight. A heavy-weight row cover (1.5 to 2 ounces per square yard) is a cozy blanket for winter crops. Expect protection down to 24°F (−4°C). These heavy-duty covers trap heat and allow you to extend the harvest of cool- and cold-season vegetables for weeks or even months. For cold-season harvesting or protecting, float covers on PVC or metal hoops so the covers don't sit on top of the plants. As with medium-weight covers, use these thick fabrics to create an overwintering tunnel for greens like baby kale, spinach, mâche, and other cold-tolerant greens.

These sturdy fabrics allow only 30 to 50 percent of light to pass through, depending on the grade and manufacturer, so don't use it for more than a few days over spring seedlings or in early autumn when ample light is required for plant growth.

SIX WAYS TO PUT ROW COVERS TO WORK IN YOUR GARDEN

Frost protection. This is the most popular use of row covers in both the spring and autumn garden. Lay them over beds, float them on hoops, or wrap them around tomato cages to shield vegetables from frost.

Heavy rain or hail protection. Sometimes spring or autumn weather goes from sun to severe in moments. A sudden downpour or brief hailstorm can easily damage or wash away just-planted seeds or seedlings. Keeping row covers

MARK YOUR COVERS

It can be hard to tell the types of row covers apart after a few uses. To simplify, when you first use your covers, write an I for insect barrier, L for lightweight, M for medium weight, or H for heavy weight in the corner of the cover with a permanent marker.

handy to toss over garden beds is an easy way to avoid or minimize severe weather damage.

Summer shading. It's true that most vegetables grow best in full sunshine, but there will be times when it's beneficial to have the ability to cast shade over parts of the garden. This is where a row cover can come in handy. Suspending the thin fabric over cool-season salad greens in late spring and early summer can delay bolting and extend the harvest by weeks. Hanging a row cover to shade the bed reduces heat and keeps the soil moist between waterings. It's also an easy way to prevent heat stress of newly transplanted seedlings.

Hold soil moisture. If holding moisture is the goal, a row cover can be laid directly on top of just-seeded beds in late spring, summer, and early autumn to raise germination rates. This is especially helpful in mid to late summer when the soil is hot and dry and it's difficult to get seeds for fall and winter crops to germinate.

Pest prevention. To prevent common garden pests like cabbage worms and flea beetles from ruining your crops use an insect barrier (see page 29). Be sure to always apply the cover *before* the pest appears. Otherwise, you may trap them beneath the cover with their favorite crop and no predators — an insect smorgasbord! Remembering to rotate your crops from one year to the next is especially important when using row cover. Many garden pests, like squash vine borers, overwinter in the soil as eggs or adults, emerging in spring. If you grow the same pest-prone vegetable in the same spot from year to year, those insects may wake up from winter slumber and find themselves under the insect barrier. You can also use insect barrier fabrics to prevent birds, deer, rabbits, and other garden pests from eating your crops.

Secure winter mulch. We winter harvest many of the root and stem vegetables in our garden. Crops like carrots, beets, celeriac, and leeks over-winter in most regions when deep mulched with shredded leaves or straw. To hold the insulating mulch in place, I use old row covers. These are placed over the mulched bed and held in place

Whether you're using poly sheeting or fabric row cover, make sure you allow extra material on the sides. You'll need to weigh it down to prevent pests from entering and to keep the sides from flapping in the wind.

Choosing the
RIGHT SIZE

Row covers come in a variety of widths and lengths. You can buy packaged, precut sheets at garden centers in sizes like 10 by 12 or 10 by 20 feet, or you can buy by the roll. Roll widths vary from 7 to 50 feet, with the widest fabrics intended for commercial use. Roll lengths range from 50 to 1000 feet.

For me, the most important factor is width. I need the sheet of row cover to be wide enough to cover my crops. And since I often float my covers on hoops, the fabric must be wide enough to accommodate the size of the hoops as well as a foot or so on either side for weighing down and securing the fabric from blowing up or away.

Insect barrier fabrics or lightweight row covers used to prevent pest damage are often left over susceptible crops for weeks or months. When sizing these fabrics, factor in plenty of slack to accommodate the growing plants.

If the cost of a roll is too steep, consider splitting the cost with gardening friends or family, or members of your garden club, community garden, or urban farm. It's usually far cheaper to purchase row cover by the roll, and you get to choose from a wider variety of sizes.

with rocks or logs. Whenever we want to harvest in winter, we move the weights, lift the row cover and push back the mulch to harvest the root crops waiting patiently beneath.

SHADE CLOTH

Shade cloth is an underappreciated tool for the vegetable gardener. Like row cover, shade cloth is a material that can be used to extend the harvest season, preserve the quality of crops, and provide shelter from inclement weather. Row cover is typically used to protect in cold weather, whereas shade cloth is used to protect crops in hot weather. Although it's true that a row cover can also be used to create shade in the garden, I've found shade cloth to be far more effective in reducing the ambient temperature around crops, thus delaying bolting and helping establish summer crops.

Shade cloths are a low-tech and inexpensive way to shelter vegetables in the garden, but also in garden structures like polytunnels and greenhouses. They're made of black or dark green polypropylene that is durable and will last for many years with proper care. The polypropylene is knitted or woven into different densities that block a certain percentage of sunlight, ranging from 5 to 95 percent. For most garden use, 30 to 50 percent shade cloth is ideal.

The type of shade cloth you buy depends on several factors including the type of plants you wish to shelter, the reason for the shade, and your geographic location. For example, northern gardeners like me who want to delay bolting of spring salad greens will find that 30 percent shade cloth is enough to slow flowering. But gardeners in Southern regions that experience more intense heat should use 50 percent shade cloth for the same job.

Shade cloth is available in precut sizes (most commonly in 6-by-12-foot sheets), rolls, or by the running foot. Precut shade cloth may come with grommets for easy hanging in structures or on top of polytunnels or greenhouses. No grommets? No problem! You can buy clip-on grommets from a building supply store.

If you don't want to make your own shade cloth support, several garden supply companies sell shade cloth tunnels that are around 2 to 3 feet wide and 10 to 12 feet long. These can be used to top garden beds in late spring or summer.

Shade cloth also makes an effective wind screen. Float it on hoops over a bed to protect tender seedlings from gusty winds.

Five Uses for
SHADE CLOTH
in Your Vegetable Garden

▶ **Delay bolting.** To delay bolting and prevent bitterness of cool-season greens like spinach and arugula in the late spring and early summer garden.

▶ **Harden off seedlings.** To create a hardening-off area for pots and flats of seedlings raised under grow lights or in a sunny window. Also, to protect young seedlings from inclement weather like hail or heavy rain.

▶ **Retain soil moisture.** To provide shade that helps the soil retain moisture, which reduces stress on transplants.

▶ **Layer for shade in larger structures.** To offer relief from sun in structures like polytunnels, greenhouses, or domes while growing leafy greens and other cool-weather crops like broccoli and cabbage in spring, summer, and autumn.

▶ **Protect against sunscald.** To protect vegetables like tomatoes, eggplant, and peppers that can be prone to sunscald or wilting in intense heat in hot-summer regions.

Securing Your Cover

I learned early on that you need to secure the covers of mini hoop tunnels well. Even the heaviest of covers can quickly blow off a bed or mini tunnel in gusty conditions. There are many ways to keep covers in place. Garden staples, clips, and hand pegs are all available from garden supply companies, but you can also use the weight of rocks, logs, or other materials. Avoid using anything with sharp or jagged edges, especially with polyethylene. You can also buy or make C-clips or snap clips which fit snugly over the PVC or metal hoops, holding the cover to the hoops.

Of course, certain products, like staples and pegs, are meant to go through garden fabrics and into the ground to hold them in place. You can even make your own garden staples with cut up pieces of wire coat hangers. Since these products put holes in the fabric, they make the covers more prone to tears and shorten their life span.

For winter structures, covers should be closed tightly at either end. I twist the polyethylene film and tie, clip, or clamp it closed. When we harvest in winter, I open the tie and untwist the end to access the vegetables inside the tunnel.

Keeping pests out. If the cover is an insect barrier to reduce pest pressure, you'll need to completely cover the tunnel, leaving no spots where insects can enter. This means securely fastening the fabric by pinning it down with garden staples, adding weights around the perimeter, or burying the sides and ends with a layer of soil. It only works if the bugs can't get in.

Proper application time is also essential. The cover must be placed over the plants before the pest appears. Otherwise, you're just creating a perfect predator-free environment beneath the fabric for the pests to feast on your vegetables.

Growing Well in a Mini Hoop Tunnel

Knowing how to care for mini hoop tunnels and the crops they shelter will help you grow healthier vegetables.

Venting. No matter what type of structure you use, be sure to ventilate it whenever necessary; interior temperatures can rise quickly in the spring and fall, even on cloudy days. The ends of a mini hoop tunnel should be propped open to allow air to circulate. Good ventilation encourages healthy growth and reduces disease problems.

To hold the fabric or plastic open at the ends of the tunnel, use clips or clamps. As the temperature falls in the late afternoon, close up the tunnel for the night.

In winter, I don't vent my structures unless there is a brief thaw where the temperatures rise

There are many ways to secure fabric and plastic to hoops. Binder clips are great for temporary spring and autumn wire hoop tunnels, while PVC snap clips hold covers securely to my winter metal or PVC conduit tunnels.

above 40°F (4°C). I make an exception to this rule for the low wire mini hoop tunnels that are in my winter polytunnel. I erect these over the beds to provide additional insulation during the deep freeze of winter, which typically runs from early January to mid-February. If we have a stretch of mild days, I remove these interior covers so the crops can enjoy maximum sunlight and air circulation.

Watering. Mini hoop tunnels covered in row cover or shade cloth won't need extra watering, as rain or irrigation water passes through the cover. But polyethylene-covered tunnels prevent water from reaching the plants, so you'll need to get the hose out.

Unless you remove the cover, it's hard to water a mini hoop tunnel with a watering can. I prefer to use a hose with a watering wand so that I can irrigate evenly. Try to water in the morning, so that the plants have time to dry before nightfall. Wet foliage speeds up the spread of diseases.

Watering is a regular chore in spring, summer, and, autumn, but I don't water in winter when transpiration is low and the crops are not actively growing.

Adding extra winter protection. In early autumn when frost threatens, I erect the hoops for mini tunnels over my raised beds. For the first few weeks, I rely on row covers to protect the vegetables from frost and cold weather. But as the autumn weather turns colder, I top the row covers with polyethylene for added protection.

Keeping Covers Clean

It doesn't take long for a pristine white row cover to go from fab to drab in the garden. It's a fact: row covers work hard and they get dirty. That dirt blocks light from passing through the covers, but it's also abrasive and can damage the covers, shortening their life span. It pays to keep on top of cleaning row covers.

The easiest and fastest way to clean row covers is to hang them up and hose them off. I hang them on my garden fence or on a clothesline and give them a good shower with my hose nozzle set to

Vent the ends of mini hoop tunnels when the temperature is above 40°F (4°C) to prevent heat buildup.

When row covers get dirty, they block light. Wash them on the gentle cycle of a washing machine or spray them with a hard jet of water and hang them to dry.

MAKE IT LAST

ROW COVER. The life span of row cover depends on how you're using it and its grade. Super lightweight insect barrier is more prone to rips and tears, so handle with care. Thick, heavy-duty row cover is pretty sturdy. If you're covering hoops with fabric row cover, the types of clips or clamps you use to secure the material also come into play. Certain clips, like those with sharp edges, can easily tear fabric, decreasing its life span. I find the edges of my DIY C-clips need to be sanded down before they're put to use with tear-prone fabrics. The low tunnel clips I've bought from garden and greenhouse supply stores have been less likely to tear covers.

SHADE CLOTH. Shade cloth is very sturdy and should last at least 10 to 12 years.

POLYETHYLENE FOR MINI HOOPS. I like to recycle plastic wherever possible, so often a portion of the polyethylene film from my poly-tunnel is reused to cover mini hoop tunnels. I can get a season or two from this recycled film, but if I'm using new polyethylene for my mini tunnels, it lasts for 5 to 6 years.

"jet." You can also wash row covers in the washing machine on the delicate cycle. After they're clean, hang them to dry and then fold and store until their next use.

Because shade cloth is often hung over plants, I find it rarely gets as dirty as row covers. If it does need a wash, hang it on a clothesline or lay it on the lawn to hose it off. Hang to dry in the sun and then reuse or fold and store.

Once mini tunnels come down for the season, I lay polyethylene sheets on the lawn, hose them off, and scrub them with water or water mixed with a mild detergent. A long-handled sponge mop is ideal for this task.

COLD FRAMES

A cold frame is simply a box with a clear top. It's one of the best crop covers for gardeners who are new to growing under cover, because it's easy to use, easy to care for, and it can shelter a wide variety of food crops in spring, autumn, and winter. The cold frame creates a microclimate around plants by trapping heat from the sun; it also protects plants from drying winter winds and inclement weather like hail and heavy rain.

I've been using cold frames for many years to grow cold-tolerant vegetables and herbs in autumn, winter, and early spring, but other gardeners put them to work starting spring seedlings for their vegetable beds. Cold frames are also a convenient spot to harden off seedlings that were grown indoors, giving them a chance to adjust to outdoor light before they're moved to their permanent place in the garden.

The Best Site

Maximum sun exposure in winter is essential for growing veggies in a cold frame through the winter, so find a south-facing site for your frame. We placed ours in a sloped, south-facing spot in the garden where the snow melts first each spring. You'll want to avoid low-lying areas where frost can pocket and drainage may be poor. If you plan on harvesting from your frames quite often, it's nice to have them in a spot that is convenient to the house and easily accessible for venting and watering.

Cold frames can be freestanding or tucked up against a structure like your home, shed, or greenhouse. Having the north side of the frames against a structure helps protect it from cold north winds. If they're freestanding, they can be placed on top of the soil or buried several inches below grade, to further insulate from winter winds and cold temperatures.

Whether homemade or purchased, cold frames can be made from a variety of materials. Those with wooden frames are ideal for winter harvesting, while polycarbonate frames stretch the spring and fall planting season by 2 months or more.

Cold Frame Kits

Premade cold frames or kits are available in a range of sizes from garden supply stores. If you decide to buy one instead of making it yourself, do some research first and read online reviews or check out the product in person if the store is nearby. Some of the higher-end cold frame kits have boxes made from redwood or cedar with twin-wall polycarbonate glazing. Others have aluminum frames with a polycarbonate box and top. There are even some nifty frames with sliding tops, double boxes, and automated vent openers. These would all look pretty sharp in a garden, but if you're on a budget I'm not sure you'll harvest enough salad greens to justify the cost. They are pretty pricey and generally quite small, often just 2 by 3 feet.

Choosing Materials for DIY

There are a variety of options for building your own cold frame, depending on how permanent you'd like the structure to be, what materials you have available, and how skilled you are at carpentry. Are you looking for a quick and easy cold frame, or something more durable, to use season after season? Here's a guide to all the materials you might consider using.

THE BOX

Cold frame boxes can be built from a variety of materials, including wood, bricks or cinder blocks, and polycarbonate. For a super quick and easy cold frame box, use straw bales.

Wood. Wood is my cold frame box material of choice due to its insulating properties, strength, reasonable cost, and how easy it is to source and work with. Plus, using a rot-resistant lumber like hemlock results in cold frames that last 8 to 10 years in my garden. Wood also makes it easy to build boxes with a 10- to 15-degree angle for maximum light exposure.

Bricks or masonry blocks. Cold frame boxes can also be built from bricks or cinder blocks, which do double duty by absorbing heat. It's a lot of work to build an angled cold frame box from bricks and cinder blocks, however, and requires special tools. For a no-fuss temporary frame, make the box from bricks or cinder blocks piled 12 to 18 inches high. If you want to make it a permanent structure, add mortar between the layers.

Polycarbonate. Another option for cold frame boxes is to build or buy one with a transparent box made from polycarbonate. Polycarbonate cold frames look pretty sleek in the garden with aluminum or wood frames holding the polycarbonate in place. They do allow more light to enter the structure than a cold frame with a wooden box, but they don't hold heat or insulate as well in cold weather. They're also not as strong as wood in heavy snow or strong winds. I've used them for sheltering crops in spring and autumn, but their winter performance doesn't come close to a cold frame with a wooden box. To successfully harvest winter crops in a polycarbonate-sided frame, I've had to stick to the hardiest vegetables like kale, tatsoi, and mâche. I've also had success with root crops like winter carrots as long as the frame was also filled with shredded leaves for extra insulation.

Straw bales. There is one other material that can be used to make a fast, inexpensive cold frame box with no carpentry skills required — straw

WHICH WOOD to Choose?

Wood is the most popular material for cold frame boxes, but eventually it rots. And as much as I'd love my cold frames to last forever, they do have a limited life span. Yet not all wood rots at the same rate, so do a bit of homework before you choose which type of lumber to use for your cold frame box.

TREATED LUMBER. This is wood that has been preserved with chemicals to delay decay. There is a lot of concern about using treated wood for garden beds and frames and having toxic chemicals and heavy metals leach into the soil. If you want to meet organic standards, avoid pressure-treated wood.

SOFTWOOD VERSUS HARDWOOD. I'm often asked if hardwood lumber is better than softwood for cold frames. Slow-growing hardwoods like black locust and white oak are more resistant to decay than softwoods like pine and spruce, but I've been using hemlock, a softwood, for my garden structures for almost two decades.

ROT-RESISTANT HEMLOCK. Our hemlock is locally grown and untreated and I've used it to build dozens of raised garden beds and cold frame boxes.

It's a dense wood that lasts at least 8 to 10 years in our garden. It's also reasonably priced and ages to a nice gray patina. Cedar is another rot-resistant softwood that is popular for garden projects and structures, but is far more expensive than hemlock.

THICKER LASTS LONGER. Thicker boards take longer to rot, so extend the life span of your frames with wood at least 2 inches thick. I use 2-inch hemlock planks for our cold frames, raised beds, and other garden structures.

CHOOSE LOCAL OR FSC. Before you decide on a wood, do a little research and ask questions at your local lumber mill or market. Choosing a local wood lessens the environmental impact and so does buying lumber that is Forest Stewardship Council (FSC) approved.

bales! I often pile straw bales around tall crops like kale, collards, Italian parsley, or leeks in late autumn and top the bales with a piece of polycarbonate for an instant cold frame. Avoid using polyethylene sheeting unless it's secured to a frame; it will catch rain water or snowmelt and sag, falling in on top of the crops or freezing into a skating rink on top of your straw bales. Not convenient when you want to harvest!

THE SASH

A cold frame top is where the magic happens. A transparent cover, also called a light or a sash, allows light to enter the structure and traps heat, creating a protected space for your crops. I've used many materials as cold frame tops and also visited dozens of gardens where cold frames have been used to extend the harvest. Here are some suggestions for what to use to top your cold frame:

Polycarbonate. Twin-wall polycarbonate is my preferred material for a cold frame sash. It's easy to cut to size, long-lasting, lightweight, more insulating than glass, and boasts an impact resistance that is 200 times that of glass. I've even caught a deer standing on the polycarbonate top of my closed winter cold frame and the top remained intact. Talk about impressive!

The polycarbonate I use is 8 mil thick and permits 82 percent of light to enter my cold frames. I purchase it at my local greenhouse supply store. You'll find thinner twin-wall products available at home improvement stores, but keep in mind they won't be as strong or as insulating.

Old windows. One of the easiest ways to top a cold frame is by using an old window, with the box built to match the size of the window frame. Because glass breaks so easily, however — and the last thing you want is to be finding glass shards in your soil for years to come! — I recommend removing the glass and attaching a sheet of plastic (either polycarbonate or polyethylene) to the frame. If you're worried about the potential of lead paint flaking into the soil, it's best to avoid using any window you suspect is from before

1978 (when lead paint was banned in the United States) and keep an eye out for vinyl replacement windows to use instead.

An alternative to an old window is an old shower door. They're made with tempered glass, and they allow you to build a much bigger cold frame.

These hemlock cold frames are sunk into the ground to boost insulation for winter harvesting. For aboveground cold frames, you can mound leaves or evergreen boughs around the frames in late fall to provide added protection.

TURN AN OLD WINDOW INTO A SAFE COLD FRAME SASH

To turn a window into a cold frame sash, you'll want to start by removing the glass. Place the window on a tarp. Put on some protective goggles and gloves, use a hammer to shatter the glass, and then sand down the wood frame to remove all shards. Once the glass is gone, cut two pieces of greenhouse-grade polyethylene large enough to cover the entire wooden frame of the window. Staple a sheet to the top and bottom of the frame, using heavy-duty staples. Voila! A simple, but effective, double-layered cold frame top that is shatterproof and insulating.

A cold frame can be built to fit the available space. Our 3-by-6-foot frames offer plenty of growing space, but still allow us to plant, tend, and harvest comfortably from the one open side. Larger frames have sizeable tops that can be heavy to lift and support for venting.

Sized for Accessibility

Our cold frames are 3 feet wide by 6 feet long, which I think is the perfect size. They are large enough to offer plenty of growing space but small enough that they can be planted, weeded, maintained, and harvested easily from one side — no reaching or stretching required. Creating larger cold frames means the top will be considerably heavier and more difficult to lift and prop open securely. You've probably heard the old adage, "The bigger they are, the harder they fall." This certainly holds true for cold frames; I've seen the top of a heavy 4-by-8-foot cold frame slam shut in a high wind, splintering the wooden frame of the sash. The polycarbonate sheet was unaffected, but a new frame had to be made.

Build a frame tall enough to accommodate the crops you want to grow. You don't want plants touching the top of the sash in winter, as the cold material will damage the plant tissues. The back of our boxes measure 18 inches tall and the front of the frames are 12 inches tall. In our frames, we tend to plant low-growing to medium-size crops like lettuce, baby chard, spinach, arugula, endive, carrots, beets, Asian greens, mâche, claytonia, and parsley. We protect taller vegetables with mini hoop tunnels or grow them in our polytunnel.

THE IDEAL SASH ANGLE

Traditional advice suggests that the slope of the sash from back to front should be 10 to 15 percent, or around one to 2 inches per foot. Some gardeners prefer much steeper angles. We have a 6-inch differential from the back to the front of our frames — 18 to 12 inches — which has worked very well. We plant taller crops like dwarf kales, claytonia, parsley, and scallions at the back of our frame, and shorter mâche, leaf lettuce, and tatsoi along the front.

Many manufactured planters now come with removable tops to support insect netting or plastic covers for season extension.

Adding a Removable Cold Frame Top to a Raised Bed or Planter

Another option is to create a removable cold frame top for a raised bed, which can be switched out from season to season or whenever the need for protection changes. For example, you could cover the bed with an insect barrier frame top for spring and summer, a row cover frame top for early spring and autumn, or a polycarbonate frame top for winter harvesting. If your crops are threatened by deer or rabbits, chicken wire frame tops are a great choice. When using removable cold frame tops, it's a good idea to stick to smaller 2-by-4- or 3-by-5-foot beds rather than typical 4-by-8-foot beds, as larger frame tops can be heavy and cumbersome to maneuver.

Many companies that sell elevated planter kits — ideal for growing food on decks and patios — offer additional accessories like protective frames, cold frames, or mini tunnels. These devices are custom made to fit their products, so if you have an elevated planter, check with the company to see if they offer any covers that can be used to protect or extend the harvest.

COLD FRAMES + ROW COVERS

Drape a row cover over an opened cold frame to keep insects, deer, and other pests out in spring and autumn. A row cover will also boost germination rates in the cold frame for crops like carrots, kohlrabi, cabbage, and kale that are sown in midsummer for autumn and winter harvesting by providing shade. Remove the cover once the seeds have germinated.

A Use for Every Season

There are so many ways to use a cold frame to grow more food! I love using mine to harvest vegetables and herbs all winter long, but they can also be used in spring, summer, and autumn.

In spring, you can use the protective cover of a cold frame to start seedlings for the garden. In early to midsummer, you might grow cover crops in your cold frames to enrich the soil or plant heat-loving vegetables. In midsummer, you can start sowing seeds for fall and winter carrots inside your cold frame. A few weeks later, beets and winter radishes can be sown, and eventually seeds for a variety of cold-tolerant salad greens.

Water Well and Vent Often

To optimize plant growth, pay attention to the two most important tasks in cold frame care: watering and venting.

Watering. Consistent moisture is crucial for healthy plant growth. Watering crops grown under the protective cover of a cold frame is not much different from growing them in the open garden. In spring, summer, and early autumn, you can open the tops (or remove them in summer) to let the rain water your crops. In late autumn, winter, and early spring — or whenever the top is closed — you'll need to supply the water.

In spring and autumn, water cold frames early in the day to give the plants a chance to dry off before the frames are shut for the night. Wet foliage at night can promote the spread of disease. If a heavy downpour is in the forecast, close your frames to protect the soil and plants from a severe drenching.

In my Zone 5B garden, I don't water in the winter. Generally, the last watering of the cold frame crops is in early December before the ground freezes. After that, the frames are shut for the winter, except when we harvest. Gardeners in warmer regions may need to get out their hose or watering can occasionally in winter to irrigate cold frame crops.

How often to water will depend on the season, temperature, the types of crops, and what growth stage the crops are in. The warm, protected environment of a cold frame, especially if not well vented, speeds up evaporation. You may

Provide regular, consistent moisture for cold frame crops, especially as temperatures rise in spring.

Vent cold frames often to prevent heat buildup and encourage good air circulation.

need to water your crops more often than in the garden. Full-grown plants will need more water than seedlings and young plants will. Don't wait for plants to wilt to pull out the hose. Ideally, you want to keep soil consistently moist, but not soaked. A daily check in spring, summer, and fall, and every week or so in winter is the best way to gauge if watering is necessary.

Venting. If there is one cold frame skill to master, it's learning when to vent. Here's my rule: if you're wondering if it's warm enough to vent, chances are that it is. I always err on the side of venting. Growing plants too warm promotes soft, green growth which is easily damaged when the temperature plunges. A little tough love goes a long way to promoting healthy, hardy vegetables.

Tucking a thermometer inside your cold frame is a good way to keep track of temperature. Even better is buying a digital thermometer or weather station that hooks up to your phone so that you can monitor your frames (or polytunnel, greenhouse, dome) from the comfort of your home.

Or you can go low-tech and just look at the weather. When the outside temperature is above 40°F (4°C), it's time to crack open the top. Even opening the cover a few inches will prevent heat from building up. If the daytime temperature is expected to rise above 50°F (10°C), I prop the

cover wide open. By late afternoon, an hour or so before the sun sets, I close the top for the night.

To prop the frames open, I use a rock, log, or stick, but you can also buy or make a notched prop stick to hold the frame open. If you have the budget, you can buy a temperature-responsive automatic vent opener. Available at many garden supply stores, this handy device automatically vents the structure once a set temperature has been reached.

Cold Frame Soil Care

I treat my cold frames like any other garden bed, working amendments like compost, aged manure, chopped leaves, and composted seaweed into the soil between successive crops. Because our soil is acidic, I also apply dolomitic limestone annually to maintain a pH in the 6.0 to 7.0 range for healthy plant growth. An occasional soil test helps me keep an eye on the pH, organic matter, and levels of major nutrients.

Green manure crops. I've also learned that an easy way to boost soil organic matter and improve the structure of the soil is with fast-growing green manure crops. Because our frames are usually in production from mid-July to early May, I have to fit the green manure crop in during the downtime. Once the last of the winter and early spring vegetables have been cleaned out in early May, I sow seed for a fast-growing green manure crop, like buckwheat. I dig the plants under before they have a chance to flower and set seeds, and by mid-July, I'm back to seeding carrots for winter harvesting.

Cold Frame Maintenance

Removing snow. Since winter structures need to capture sunlight to be effective, snow must be removed. I always try to get out soon after a storm before the snow has a chance to form an icy layer on top of the sash. Using a broom or plastic shovel is best, as metal shovels can scratch polycarbonate tops.

There is an exception to snow clearing, however. During the period of deep winter freeze when the temperatures are extremely cold — typically mid-January to mid-February — I often let a blanket of snow accumulate on the top of the frames. This provides added insulation to any root or salad crops that we're still harvesting. Once late February arrives, all is cleared out and fresh crops replanted for an extra-early spring harvest.

Cleaning tops. Putting a little annual TLC into your cold frame can help it last longer and be more efficient. One of most important tasks is to keep the top clean. Dusty, dirty, or leaf-covered cold frame sashes reduce the amount of light reaching your plants and make the cold frame less effective.

In spring and autumn, or anytime the top looks dirty, grab a soft rag and a container of water mixed with a mild detergent or a water-vinegar solution. Dampen or spray the cloth and wipe away any dirt. Avoid spraying any of the cleaning solution near the plants or the soil. After cleaning, wipe the top well with clean water.

Summer storage. Once the risk of frost has passed in the spring, I remove our cold frame tops, storing them in the basement until autumn. This reduces wear and tear and extends their life span.

As I'm removing the tops for summer storage, I give the frames a quick inspection to gauge their general condition. Are the hinges still tightly attached? Are any screws missing? Is the wood in good shape? I make any necessary repairs at that point, so that the tops are ready to go when I take them out again in mid-autumn.

SIX EASY WAYS
to Boost the Insulation of Your Cold Frame

▶ **Line it.** Line the inside of the cold frame with 1-inch-thick foam insulation board.

▶ **Add thermal collectors.** Thermal collectors, like 2-liter soda bottles, can help retain heat. Paint them black to increase heat absorption and put them in the back of your cold frame.

▶ **Surround it.** If your cold frame isn't next to a building, place some straw bales, evergreen boughs, or bags of leaves along the north side in winter.

▶ **Bury it.** Bury your cold frames or mound a few inches of soil or bark mulch around the perimeter of the box to prevent cold air from seeping in.

▶ **Seal it.** Attach weather stripping to the inside edge of the cold frame top. This improves the seal, trapping more heat inside the frame.

▶ **Cover it.** On frigid nights, toss an old blanket, carpet, or other insulating material over the top of the frame. Remove it in the morning to allow light to enter the frame.

Dirty cold frame tops block light and reduce the effectiveness of your frames. Wipe them clean with a soft, damp cloth or sponge if they become grimy or dusty.

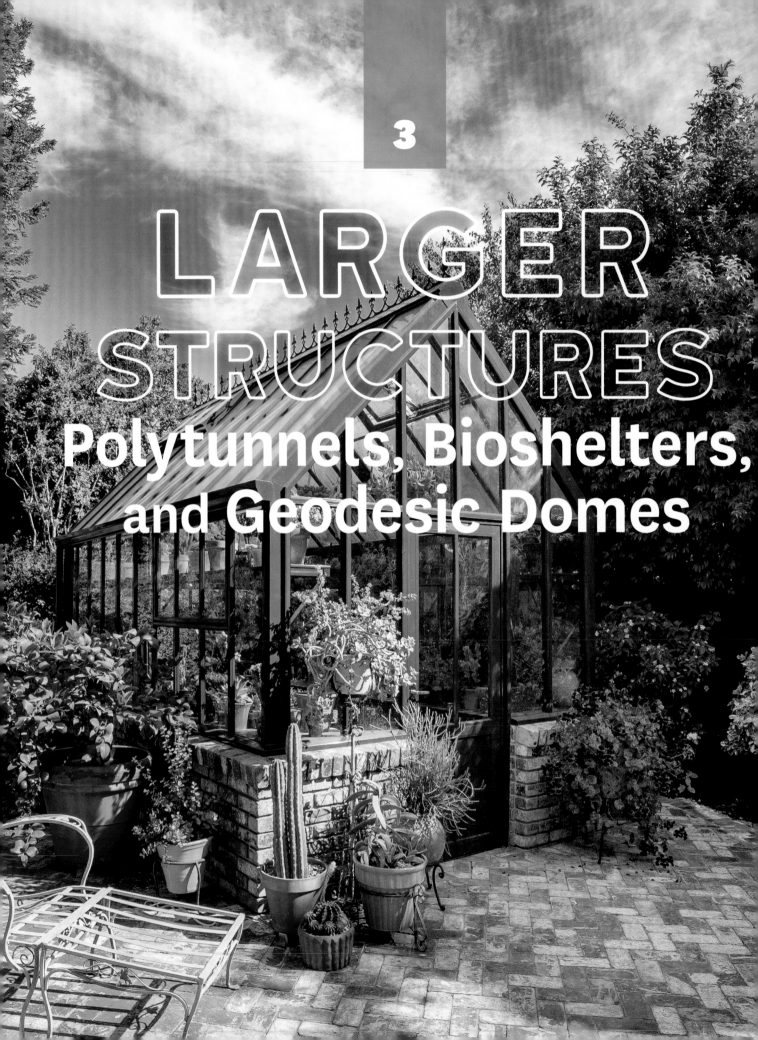

3

LARGER STRUCTURES

Polytunnels, Bioshelters, and Geodesic Domes

I think every gardener dreams of having a sheltered space like a greenhouse or polytunnel where they can garden no matter the weather. Over the years, I've grown food in various types and sizes of walk-in structures. Some, like my current polytunnel, were strong and winter-sturdy for year-round growing. Others, like the flimsy clear tent that was my first greenhouse, were more seasonal and temporary.

And although I love my cold frames and mini hoop tunnels, I am *in love* with my polytunnel. Larger structures offer many advantages over smaller garden devices. Increased growing space is the most obvious benefit of course, but growing in larger structures also results in healthier and higher-yielding plants. It's also a great way to extend the harvest season of heat-loving vegetables like tomatoes, cucumbers and melons, as well as harvest cool and cold-hardy crops all winter long. I also love that my polytunnel is a bubble in the middle of the garden, offering a comfortable space to work. It has quickly become my favorite place to hang out — no matter the season.

Before you break ground on a site for a permanent structure like a polytunnel or greenhouse, put some thought into picking a good site. It should be level, offer plenty of direct sun, and be conveniently located to your home or garden.

CHOOSING THE RIGHT SITE

There are four main site conditions to consider when locating a greenhouse, dome, or polytunnel: light, accessibility, drainage, and wind.

Lots of Light

Light is essential for healthy plant growth, so your site should offer plenty of sunlight. Most crops need at least 6 to 8 hours of full sun in order to grow well. Light is also important in winter when solar energy warms your cold-season structures. An east-west orientation is generally recommended for greenhouses and polytunnels, especially those to be used for winter harvesting. This means that the ridge of the structure runs east-west. In winter, this permits the low track of the sun to hit the south side of the structure, letting in the most light possible. Avoid placing cold frames, greenhouse, domes, or polytunnels in the shade of tall trees, hedges, fences, buildings, and other structures.

Finding just the right spot in a small yard or urban setting can sometimes be a challenge; you might not have a lot of options. We built our polytunnel in the one spot on our property that had the space to accommodate it; it runs southeast-northwest. Like us, you may only have one site that fits your desired structure, but if you have some choice, take a few minutes and look at each potential location to determine the best option.

Plan for Easy Access

I like being in my polytunnel. I like harvesting from my cold frames. And I like keeping an eye on my crops so that I can ensure they are growing well. That means I want my protected garden beds to be sited where they'll be close at hand. Many are in the garden: row covers, mini hoop tunnels, shade cloth, cloches. But my polytunnel and cold frames are just outside the garden fence

CHECK YOUR BYLAWS AND BE NEIGHBORLY

Small, temporary structures like cold frames and mini hoop tunnels don't need building permits, but larger, more permanent ones like polytunnels or greenhouses may require one. Check your local bylaws before you start the building process, being sure to check on any size limitations. Certain regions require a permit for greenhouses over 200 square feet, for example. If your neighborhood has a homeowners' association, I'd also check to see if there are any restrictions on garden structures like polytunnels.

In the interest of neighborly courtesy, give your neighbors a heads-up if your plans include a walk-in structure, especially if it will be sited near the property line. Greenhouses are often considered ornamental, but many people still cling to the belief that a polytunnel is an eyesore. I strongly disagree, but it's up to you to talk it through with your neighbors. Some advice from me: an occasional basket of homegrown veggies goes a long way toward keeping neighbors happy.

and having them a mere 30-second walk from my back door makes life easy when I need to water, ventilate, or close cold frame covers.

CLOSE TO A HOSE BIB AND ELECTRICAL OUTLET

It's a good idea to site your structures close to water and electrical sources. In a small garden, it may not seem like much work to water with a watering can, but when you factor in extra structures, especially those that are sheltered from the rain, it makes sense to have a water source nearby for a hose hookup. If you decide to hook up a faucet or irrigation system inside a greenhouse or polytunnel, you'll need to run a water line deep enough so that it is below the frost line.

An electrical outlet is also a bonus. I don't use extra heat or lights for the plants in my structures, but there are times I want to work after dark (especially in late autumn when darkness falls in late afternoon) and I end up putting on a head light. Oh, the glamorous life of a gardener!

Electricity isn't just about light, however. If you want to heat your structure, install a ventilation system, or use an inflation fan to create an air-insulated double layer of polyethylene to reduce heat loss, you'll need an electrical source.

Our polytunnel was built close to an outside tap to allow for easy irrigation.

When planning raised beds in a garden or walk-in structure, leave space between beds to allow easy access for planting, harvesting, and soil care — even in late summer, when beds are overflowing!

MULCHED PATHS AND WIDE CLEARANCE

Having mud-free access to garden structures is also nice. I lay bark mulch between my raised garden beds and around structures like my cold frames and polytunnel. I replace the bark mulch every 3 years, but you can also use a permanent mulch or pathway material like pea gravel, bricks, or flagstones.

Finally, remember that a garden structure like a dome, greenhouse, or polytunnel isn't a "build-and-forget-about-it" project. Leave at least a yard or two of clearance around the entire perimeter to give you space for repairs, rolling up sides, replacing polyethylene, cleaning, shoveling snow, and other chores.

Good Drainage

Good drainage is essential around a garden structure — big or small. Rain and irrigation water should drain away from mini hoop tunnels, cold frames, greenhouses, or polytunnels as quickly as possible. Sitting water not only quickly kills plants but can also rot wooden foundations and make a mess in large, walk-in structures. If your site is flat, drainage pipes can be added to draw water away.

I've gardened for decades with my gardens always sited on ground with a slight slope. It was only when we built our polytunnel that I witnessed the importance of good drainage firsthand. That first rainstorm I sat in the tunnel and watched the hard rain pour off the plastic and onto the surrounding area. I was grateful that our structure was sited on land that is slightly sloped so that the water drained downhill. Otherwise, the perimeter of the structure would have been a muddy mess.

Avoid Windy Sites

A gentle breeze is perfect for ventilating a structure, especially from late spring through mid-autumn. But every time a hurricane, strong wind storm, or nor'easter rips through my region, I cross my fingers that our polytunnel survives intact. Strong winds can shred polyethylene coverings, and even twist metal hoops if the gusts are strong enough. Having a sturdy, well-anchored, and well-covered structure will reduce damage, but so will siting it in a sheltered area. Figure out where your prevailing wind is coming from and aim to place your structure near a windbreak, tree line (but far enough away that it won't shade or damage the structure if a branch should fall), slight slope, or any spot that offers some shelter. Avoid hilltops if possible, as they tend to be gusty. Also, avoid the bottom of a hill where frost can pocket and water runoff may cause flooding.

In an open location, a hedge or windbreak can be planted or installed to reduce wind damage. Just plant or erect it far enough away that it won't shade the structure. The downslope of a slight knoll is also ideal, offering excellent drainage and shelter from wind.

/ If possible, avoid low-lying sites or those at the bottom of a hill. Water can pool there, and since cold air drains downhill, they can also be frost pockets.

PREPPING THE SITE

A well-prepared site makes construction easier and helps prevent future drainage problems, persistent weeds, and other issues that can cost time or money down the road. The amount of site preparation depends on what type of structure is being used. Small, temporary structures like mini hoop tunnels, whether covered in row covers, shade cloth, or polyethylene, are generally placed over existing garden beds and don't require much prep. More permanent structures, like greenhouses, domes, and polytunnels, on the other hand, need a site that has been properly prepared before construction begins.

Level it. Having a site with a minor slope is beneficial in helping water drain away, but you'll still want the grade to be as level as possible, to make installation easier. In the case of a severe slope, it's best to terrace the land before building the greenhouse, dome, or polytunnel. Site preparation typically involves removing existing vegetation and then breaking up the ground by hand, with a rototiller or, in the case of larger structures, a tractor. We rented a mini tractor to level our polytunnel site and had a friend bring his laser level to help us ensure our site was flat.

Clear away debris. We built our polytunnel where there was an old raised-bed vegetable garden that had been out of production for a few years, so it was a jungle of perennial weeds. I removed as many of these as possible with my

garden fork and, in the process, dug up a surprising pile of dead tree roots and rocks. It was a lot of work, but it was worth it! If I wasn't in such a rush to get the tunnel beds prepped and planted for fall and winter, I could also have solarized the soil, a chemical-free way to eliminate future weeds by killing the seed bank of weed seeds in the soil. Solarizing also eliminates soil pests like nematodes and pathogens like verticillium wilt.

The best time to solarize the soil is once the site is prepped, weeded, and levelled. Keep in mind that solarizing, which is a lot like cooking the soil, takes around 2 months. In my northern climate, the best time to solarize is in early summer when the sun is most intense. Begin by giving the site a good watering and then cover it with a sheet of clear polyethylene. Hold the edges of the plastic securely in place with rocks, sandbags, logs, or other weights. Ideally the top few inches of the soil should reach 140°F (60°C). After about 2 months, remove the plastic and continue with polytunnel or bed building. Even after your structure is built, there may be times you wish to solarize your polytunnel soil to reduce weeds, pests, or soilborne diseases. You can do this by laying the clear plastic inside the tunnel.

Scout the surrounding site. Site preparation should include inspecting the surrounding area for trees, shrubs, hedges, and buildings, assessing their health and whether they are structurally sound. You don't want a tree, branch, or piece of siding flying off in a windstorm and landing on your structure. If nearby trees are in decline or blocking light, they may need to come down before construction begins.

Water and power. Running plumbing lines and electricity is another thing to consider. It's not cheap to do, but it's better to run and bury the lines during construction than at a later date. Water is a necessity, as crops grown under the protection of a structure like a polytunnel or greenhouse need frequent watering. Our tunnel is close enough to our house that I can run a hose from the outside faucet. If yours isn't so conveniently located, you can install a water collection system like a rain barrel or you can run a water line. For our polytunnel, there was little we needed to power (besides my tea kettle!) and we knew that if we decide to eventually put in a ventilation fan, there are many solar power options available.

POLYTUNNELS AND GREENHOUSES

Picture this: It's February and there's snow on the ground. Yet as you wander into the garden and open the door to your polytunnel or greenhouse, it's noticeably warmer on the inside. You quickly shed your coat and survey the inside garden. There's a variety of salad greens, root crops like carrots and beets, and even a few aromatic parsley plants tucked into a back corner of a bed.

Sounds pretty nice, right? A lot of gardeners seem to think so, and these affordable and spacious structures are popping up in home gardens, community gardens, and urban farms across North America.

A few years ago, we added one to our garden space. I'll admit that I had my eye on a polytunnel for a long time, but we didn't have a good sunny, level spot for a structure. However, when we renovated and expanded our vegetable garden, an opportunity arose to clear and level the land beside the garden and we jumped on it.

I use wire hoops topped with row cover in my winter polytunnel. This doubling up of garden covers provides extra insulation to cold-season salad greens, scallions, parsley, and root vegetables.

What Is a Polytunnel?

Polytunnels are also called high tunnels or hoop houses and are typically made with steel frames covered with polyethylene. Like many other season extenders, a polytunnel captures solar energy, warming the interior of the structure and creating a microclimate around plants. They're usually not heated, but because they contain such a large air mass they stay warm for a longer period of time than a cold frame or mini hoop tunnel.

Seven Benefits of Growing in a Polytunnel

- ► **SEASON EXTENSION.** Stretch the harvest into fall and winter and get started earlier in the spring.

- ► **CREATING A MICROCLIMATE.** A polytunnel is an excellent way for gardeners with short summers to provide extra heat to crops like tomatoes, melons, cucumbers, eggplant, and peppers.

- ► **PROTECTION FROM EXTREME WEATHER.** Not only does a polytunnel capture heat, it also protects crops from hail, heavy winds, and other extreme weather.

- ► **A PLACE FOR PROPAGATION.** A polytunnel provides the perfect environment for starting your own seedlings to plug into the garden, containers, or the polytunnel.

- ► **DETERRING PESTS.** In my garden, I constantly battle deer, groundhogs, and the occasional rabbit, but my polytunnel crops are safe from these common pests. If insect screening is installed, you can also exclude insect pests like cabbage worms and cucumber beetles.

- ► **PREVENT DISEASE.** Wet foliage is an invitation to disease. Growing under the protective cover of a polytunnel reduces the occurrence and spread of many common plant diseases.

- ► **GARDEN IN COMFORT.** On all but the coldest days of winter, our polytunnel is a place to harvest, seed, transplant, repot, or plan the garden in relative comfort. We even added a sitting area to the back corner because we love spending time out there.

Higher-Quality CROPS

Not only does my polytunnel extend the harvest of warm-season vegetables, it also helps me grow higher-quality crops. For example, as summer turns to autumn, the cherry tomato plants in the open garden decline in production, but also in fruit quality. The autumn-produced tomatoes are smaller with thicker skins. And because of increased rain in autumn, the tomatoes are more prone to splitting. The polytunnel, on the other hand, continues to yield a heavy crop of large, thin-skinned cherry tomatoes for another 6 to 8 weeks!

A glass greenhouse is a dream for many gardeners. If your space and budget can accommodate one, these high-quality structures can provide decades of use.

Polytunnel or Greenhouse?

Polytunnels are similar to greenhouses in that they provide a protected growing environment, but polytunnels offer several advantages over greenhouses. The biggest one is cost. Polytunnels are cheaper and quicker to build and easier to maintain and run. Plus, the wide variety of polytunnel sizes and shapes available through commercial suppliers, or the ease of DIY-ing your own, means that there is a polytunnel to fit spaces of all sizes.

When I was deciding between a greenhouse or a polytunnel, it was both the cost and the versatility of a polytunnel that made my decision. My 14-by-24-foot polytunnel was priced about the same as an 8-by-10-foot polycarbonate greenhouse, but I've got more than four times the growing space and a higher roof with strong trusses to provide support to vertical crops.

The downside of a polytunnel in a home garden is that it's less visually appealing than a greenhouse. But with the value of local food production in the headlines, more gardeners are erecting polytunnels in their suburban neighborhoods. Just be sure to check your local and neighborhood bylaws before you break ground.

Types of Polytunnels

Once we made the decision to add a polytunnel to our garden, I got serious about researching the best type of structure to fit our needs. I wanted to have a walk-in space to harvest from year-round. I wanted to extend the growing season in spring and fall for heat-loving crops like tomatoes, peppers, and melons. I wanted strong horizontal trusses to support heavy vining vegetables. I wanted a structure sturdy enough that it wouldn't collapse in our stormy maritime climate. And I wanted it to look nice and fit in with our garden and landscape.

That's a lot of demands, and we spent hours looking at polytunnel kits trying to decide which one to choose. I knew we weren't going to DIY our own structure, as time was short and we wanted a professionally designed tunnel with strong steel hoops and all the parts and pieces included.

COMPARING POLYTUNNELS AND GREENHOUSES

POLYTUNNEL	GREENHOUSE
Cheaper per square foot	More expensive per square foot
Holds more heat	Cools quicker
Covered in polyethylene	Most have rigid coverings (polycarbonate/glass)
Curved sides take away some of the growing space	Straight sides ideal for growing as well as hanging shelves
The large space can be harder to vent	Easier to vent; more options for vent systems
Less aesthetically pleasing, especially in a home garden	Decorative and useful
Quick and easy to set up	May need to be installed by a professional

QUONSET POLYTUNNEL

GOTHIC-STYLE POLYTUNNEL

Quonset versus Gothic. Polytunnels have two basic styles: Quonset and Gothic. A Quonset-shaped tunnel is a semicircle with arches made from single pieces of steel tubing or PVC conduit. The design is simple, but the shape creates low side walls where medium or tall plants can butt up against the glazing. It also doesn't shed snow well.

Gothic polytunnels have arches made from two pieces of steel, although PVC conduit can also be used, with a peak at the center of the roof. This peak provides more headspace and straighter side walls, which means there is more room to grow crops. The peaked roof shape is also better at shedding snow than a Quonset polytunnel is.

Kits. As with most things in life, you get what you pay for when it comes to polytunnel kits. Quality varies tremendously, and tunnels run the range from flimsy junk to professional grade, with countless options in-between. The best way to figure out what to buy is to think about how you'll use the tunnel.

Most kits come with all the parts and pieces you need to build and cover the structure. Ours is a commercial-grade 14-by-24-foot polytunnel designed for strength and longevity. It was a lot of work to put up; not hard work, but there were many steps and parts that had to be followed and installed in a certain order.

We were lucky to have folks with experience helping us. Together, we were able to put up the skeleton of the structure in one day. We then took time to build and cover the gable ends. Once they were up, it was another half-day of work to cover the tunnel with polyethylene and install the roll-up sides.

Pop-up polytunnels. Many online sources offer what I call "pop-up polytunnels," which look a lot like clear tents. Way back when I was a teenager, this is what I used to start seedlings and shelter warm-weather crops. And to be honest, it did an okay job — it fit my needs at the time.

These tunnels are not very sturdy, but they are inexpensive. If you choose to buy one, bear in mind that lightweight polytunnel kits rarely come with a sturdy anchoring system and usually just sit on top of the ground. I'd recommend anchoring them with pegs or stakes so they don't blow away. Pop-up tunnels can be used from spring to autumn, but not through the winter; their arches are unlikely to support any snow load.

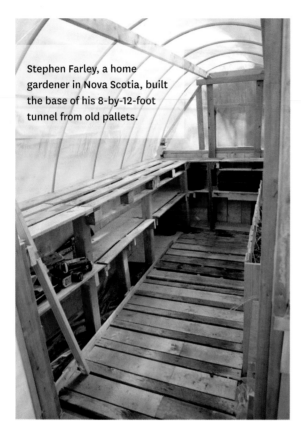

Stephen Farley, a home gardener in Nova Scotia, built the base of his 8-by-12-foot tunnel from old pallets.

DIY structures. DIY tunnels can be made with a variety of materials, but most gardeners use PVC pipes or metal hoops. A PVC structure is made with lengths of PVC conduit and connectors with the hoops anchored on steel posts or with flat-bottom brackets attached to a wooden base. PVC isn't as strong as metal, so it's better for areas that don't receive a lot of snow or for tunnels that are only used from spring through autumn (in this case, the cover should be removed in late autumn and secured again in early spring).

If you're in a region with the potential for heavy snow and plan on using the tunnel all winter long, metal hoops are your best bet. Bend your own high tunnel hoops with a metal bender. Depending on the bender you buy, you can bend either a Gothic arch or Quonset-shaped hoop, but a Gothic arch is superior for snow shedding.

Of course, gardeners can be super creative and I've seen polytunnels from a variety of materials, including an old trampoline. If you're thinking of building your own tunnel, check out some of the many tutorials and videos online for ideas, inspiration, and advice.

Cover Materials

When deciding on a cover for your greenhouse, polytunnel, or dome, you'll find a number of options available. The main covers used by home gardeners are polyethylene or polycarbonate. There are other products like acrylic panels or bubble plastics (like giant sheets of bubble wrap!) used in professional greenhouse operations, but these are less common in home gardens due to cost and availability. Each product has its advantages and disadvantages, so do some research and ask questions before you make your final decision. If you're lucky enough to live near a greenhouse supply store, go in person to check out the different options and get your questions answered. Think about what type of structure you're going to cover and how long you need your cover to last. Be sure to factor in budget, light transmission, heat retention, strength, ease of installation, and aesthetics.

POLYETHYLENE

For most unheated home polytunnels, greenhouses, and domes, especially those that are homemade, polyethylene is the cover of choice. Compared to glass or polycarbonate, it's inexpensive and easy to install, though it degrades more quickly because of light and heat. Greenhouse-grade polyethylene (available from greenhouse supply companies) costs twice as much as construction-grade plastic sheeting, but thanks to the addition of UV stabilizers, it should last four to six times as long. Our polytunnel is covered with 6 mil, 4-year greenhouse-grade polyethylene, which has 90 percent light transmission. With proper use and no unexpected severe weather events, I expect it to last for at least 6 years. If your structure is temporary and will only be in place for one to two seasons, you may choose to go with construction-grade plastic (available at hardware stores) or a 1-year greenhouse film.

Commercial greenhouse growers have even more options to consider; covers with condensation control or controlled diffusion, for example. An average home gardener likely won't use these types of products, as the benefits don't outweigh the higher price.

Although flexible polyethylene films don't last as long or wear as well as polycarbonate, it's fairly easy to repair holes and tears with greenhouse tape. I keep a roll of repair tape on hand for unexpected damage.

Home growers who want to heat their structures for winter use should consider installing a double layer of polyethylene that is inflated with a blower fan. The air cushion between the layers of film boosts insulation and reduces heat loss. It also makes a stronger, more durable cover. If you do choose to inflate the polyethylene, place the fan so that it draws in the outside air. The interior air is more humid than outside air and causes condensation between the plastic layers.

POLYCARBONATE

Polycarbonate covers are pricier than polyethylene covers, but they're stronger, longer-lasting, and offer an aesthetic that many gardeners prefer. The material is extremely lightweight and significantly stronger than glass. Also, panels made for greenhouse use are UV treated to last longer and resist yellowing. Most are guaranteed to last for 10 years.

Many greenhouses and domes available as kits come with a polycarbonate cover option, and even DIY structures can use polycarbonate as the covering. Different types and grades of polycarbonate are available at building or garden supply stores and can be attached to a DIY wooden-framed greenhouse quite easily using H-splice or end-splice aluminum extrusions.

Twin-wall. This double-layered polycarbonate sheet offers better heat retention than the single-wall type; the air space between the two sheets increases its insulating properties. It can be purchased in 4- or 8-mil-thick sheets. Some panels are rigid, while others are more flexible and can be curved to fit over the arch of a greenhouse or polytunnel. A triple-wall option is also available.

One drawback to polycarbonate is that it allows less light to pass through than glass or greenhouse polyethylene. The 8 mil twin-wall Lexan we used on the gable ends of our tunnel allows 82 percent of light to pass through.

The front gable end of my polytunnel (bottom) is made from twin wall polycarbonate. It's a strong material but it also spruces up the appearance of my poly covered tunnel (top), which is sited in my residential neighborhood.

TIPS FOR CUTTING POLYCARBONATE

► Safety is key. Be sure to take all proper precautions, including wearing protective eyewear and clothing.

► Measure and mark the cut with a grease pencil.

► Secure the polycarbonate sheet on several sawhorses, a table, or another stable surface using C-clamps.

► Measure again!

► Using a circular saw with a fine-toothed blade or a jigsaw, begin to cut through the poly-carbonate. Go slow and don't push the saw along. A circular saw is best for straight cuts, while a jigsaw is better if you need to cut the polycarbonate in a curve for the gable ends.

Many companies sell glass greenhouse kits but you can also make your own from old windows. Keep in mind that using tempered glass is safest.

GLASS

Like most gardeners, I've dreamed about a glass greenhouse: a Victorian structure with a peaked roof and decorative ridge crest where I could putter in style and comfort. Glass is a traditional material for greenhouse glazing, as well as for cloches and cold frame tops. It's also beautiful and durable, and allows up to 95 percent light transmission, more so than polyethylene and polycarbonate.

You may wonder why I call glass durable if it's so breakable. In climates where hail, heavy snow, or extreme weather aren't common (and a location where vandalism is unlikely) glass can last for decades.

Glass is also the most expensive covering for a greenhouse or garden structure. The cost of a glass greenhouse depends on several factors, including the structure style and materials, size of the panes, and whether you want single- or double-paned glass. Single-paned glass has the best light transmission at around 95 percent, but double-paned is close behind at around 90 percent. Double-paned glass is also more insulating and will retain more heat.

Six Points to Consider When
CHOOSING YOUR COVER

COST VS. LONGEVITY. Greenhouse-grade polyethylene costs about a tenth of what polycarbonate does. However, it doesn't last as long. UV-treated, greenhouse-grade polyethylene should last 5 to 6 years. Most polycarbonate products come with a 10-year warranty and typically last 12 to 15 years, barring any unexpected extreme weather events.

LIGHT TRANSMISSION. Every gardener knows that vegetables need plenty of light to grow and produce well. Be sure to take light transmission into consideration as you choose a glazing material. Refer to the table on this page for a comparison.

HEAT RETENTION. Glass makes a beautiful glazing material but it isn't as insulating as twin-wall polycarbonate. Of course, these are both expensive materials. Many commercial greenhouses use a double inflated layer of polyethylene. This type of covering does provide better heat retention than a single layer of polyethylene, but it also has increased cost and the air-inflation kit needs an electrical source. Don't forget that you can also retain heat in a structure by layering — using row cover or mini hoop tunnels inside the larger polytunnel or greenhouse.

STRENGTH. I live in an area where winter storms are frequent and I don't want to worry about my structures every time I see a snowflake. Therefore, strength is always at the top of my mind when I select a glazing material. Our polytunnel is covered in a single layer of UV-treated polyethylene, rather than a flimsier construction-grade plastic. We also built our front gable end from twin-wall polycarbonate, a strong material that can withstand gusty coastal winds.

EASE OF INSTALLATION. Certain covers require more time and skill to install. For example, the front gable end of our polytunnel is glazed in sheets of twin-wall polycarbonate, which had to be carefully measured, cut, and secured to the wooden gable frame with aluminum H-splice and end-splice (also called an end cap). The back gable, however, is covered in polyethylene, which was much easier to install.

AESTHETICS. My vegetable garden is mostly hidden from the street, but my polytunnel is visible and I wanted it to look good and fit in within the suburban landscape. This is why we invested a bit more in the front gable end and glazed it with twin-wall polycarbonate.

LIGHT TRANSMISSION OF GLAZING MATERIALS

Material	Light Transmission
Glass	90–95%
Polyethylene; UV treated	90%
Polyethylene; double layer, UV treated	80%
Polycarbonate; single layer	90%
Polycarbonate; twin-wall	80%

What Size?

Now that you know you want a polytunnel, the question is — how big? If you're going to the expense and time of building a structure, it's better to go with a slightly larger tunnel than you think you need. Why? Let's be honest, as gardeners, we can always use more growing space and you'll be surprised just how fast your tunnel fills up. Plus, you may want a spot in the polytunnel to work, putter, pot plants, or even just sit.

How much food do you realistically want to grow? I use my polytunnel to grow a wide variety of vegetables and herbs for the kitchen. I try to plant what my family will realistically be able to use, which means I am planting small amounts of each crop. I don't want a glut of scallions or radishes all ready at once. I'd rather plant modest sowings at any one time to reduce food waste.

What types of crops do you want to grow? Some vegetables occupy their space for months, while others are sown, grown, and harvested within weeks.

How much space do you have? You don't want to cram a big tunnel into a site where it doesn't make sense or where it needs to be laid out awkwardly in the overall garden design. Also, remember your tunnel should have at least 3 feet of space all around to accommodate covering, repair, and maintenance. Factor that into finding the right site.

How much time do you have to spend tending to polytunnel vegetables? My advice for a vegetable garden is to always start small. You can easily add more beds as you gain gardening skills. However, with a polytunnel, my advice is the opposite — go a little bigger than you think you need, as it's usually difficult to make a polytunnel larger after the fact. If you find that you're occasionally too busy to manage all the growing space, you can sow a cover crop in polytunnel beds to keep weeds down and build up the soil.

What's your budget? Consider going bigger. A 10-by-12-foot polytunnel gives you 50 percent more growing space than a 8-by-10-foot polytunnel, but doesn't cost 50 percent more.

What are the local zoning laws? Before you buy, make sure the size of your polytunnel isn't going to violate any ordinances.

AN OVERVIEW OF BUILDING A POLYTUNNEL

Each polytunnel kit or DIY project is unique, but most will follow these simple steps. Always refer to the manufacturer's instructions for assembly.

Prep the site. Prepare the site well in advance of building day. This means levelling the ground and removing weeds, rocks, and roots. At this point you should know how you plan on anchoring your structure, so dig holes for foundation posts or build the wooden base for the flat bottom brackets.

You want your tunnel to be square, so spend time getting the footprint right. Use the 3-4-5 triangle method to get a perfect right angle (see page 64). We used hemlock for the base of our

tunnel, which we securely pinned to the ground with 3-foot-long rebar stakes. The wood was predrilled for the stakes, which were pounded in at a 30-degree angle for a strong hold.

Take inventory. The day your polytunnel kit arrives, open all the bags and boxes, checking off the parts and pieces against the materials list. Make sure all the hardware is included and no parts are broken or damaged. If you're missing anything, inform the manufacturer immediately so they can send a replacement. This is also a good time for an initial read of the assembly instructions to familiarize yourself with the building process.

Our 14-by-24-foot polytunnel was a kit that had to be assembled. If you order a kit, take a careful inventory of all the parts and pieces when it arrives. You don't want to discover you're missing materials when you start to build, then have to stop and wait for a part to arrive.

Assemble the frame. This is the step where you invite two to four handy friends over to help with frame assembly (you supply the pizza and beer). Assembling a polytunnel isn't difficult but there can be a lot of steps and having helpers makes the process quicker. It's also much easier to lift and secure hoops when there are extra hands to help out. Have all your tools assembled and drill batteries fully charged. Make sure everyone is wearing closed-toe shoes.

First, the arches are assembled and fastened to the wooden base with flat bottom brackets (A). Then, the ridge tubes (B), cross braces, side purlins (C), corner stabilizers, and wind braces are attached to the structure. The parts and hardware assembled depends on the type and manufacturer of your structure. It's always best to stick to the specific instructions.

Build the gable ends. Once the frame of our tunnel was assembled (D), we turned to the gable ends. Each end was framed in with hemlock boards, but we decided to cover the front end in polycarbonate to give the tunnel a more ornamental appearance. This meant installing aluminum H-splice and end-splice to hold the polycarbonate in place.

Our front gable has a 36-inch-wide door (big enough to accommodate a wheelbarrow) and the back has two windows. Once the gable frames were built, the polycarbonate sheeting was installed (E) in the front and the back gable was covered with polyethylene secured with wire lock channel and Wiggle Wire.

SQUARING UP THE CORNERS

The 3-4-5 rule is based on the Pythagorean theorem and is used to make sure the corners of a structure are 90 degress, or "square." We put this simple rule to work when laying out the base of our polytunnel by measuring 3 feet out on one side of a corner and placing a mark, 4 feet out on the opposite side and placing a mark, and then running a tape measure between the two marks. If you measured your 3 and 4 foot lengths correctly, this diagonal will measure 5 feet. This handy rule will keep your structures in line — literally!

Cover the polytunnel. First, apply anti-hotspot tape (an insulating adhesive tape) to each hoop before the tunnel is covered. It extends the life span of the polyethylene by preventing friction and heat damage where the polyethylene touches the frame.

Choose a warm, wind-free day to cover the tunnel. Heat makes the polyethylene more pliable and it's easier to get a tighter fit when the weather is warm. As for wind, a gusty breeze makes it difficult — and dangerous — to install a large piece of plastic over a polytunnel frame, so keep an eye on the weather.

Polyethylene is secured to a polytunnel by trenching (a process that buries the film) or with aluminum wire lock channel and Wiggle Wire (F). We chose the second option, which was both easy to install and easy to tighten. The plastic needs to be taut, so have several helpers hold it taut while the Wiggle Wire is inserted into the wire lock channel. Wear protective eyewear when weaving the Wiggle Wire into the channel as it tends to fly about.

We also installed our roll-up sides during the covering process. The polyethylene was slid over the frame so that it was even. We left an excess foot or so of plastic at the bottom of each side. The steel tubes were placed on the excess plastic (G) and we rolled the tube up tightly, several times, securing it with clips (H). A hand lever was installed at the end of each tube for easy roll up (I).

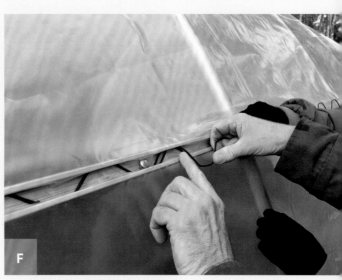

F

Polytunnel Extras

A sitting area. You may want to include a chair or bench in your own tunnel, or even a small table for potting and seeding. Once our polytunnel was built and the beds in place, we turned the back corner into a brick sitting area. I knew I was giving up valuable growing space, but one of the reasons I went with a 14-by-24-foot polytunnel was so that there would be room for a spot to read, plan, and relax, as well as enjoy the occasional mug of tea.

Accessible storage. I keep a basket near the entrance of my tunnel for all my garden gear — pruners, trowels, gloves, hat, sunscreen, waterproof markers, plant labels, twine, and tomato clips. If you have the space inside or outside the tunnel, keep a large, plastic lidded container for storing pots, trays, cell packs, row covers, and so on. Having specific storage for items like these helps keep the tunnel tidy. I also use the horizontal boards of the back gable end to hold a waterproof radio and spare pots.

Work space. Consider including a potting bench or table, and stash an extra bag of aged manure and potting mix down below. The bench can also hold shallow trays of microgreens in spring and fall. These go from seed to harvest in just 2 to 3 weeks and don't take up much space.

A step stool. A small, sturdy step stool is handy for when you're training and pruning vining crops.

When I designed the interior space of my polytunnel, I knew I wanted to include a small sitting area. My bright blue bench has become my favorite spot for reading, writing, and enjoying a cup of tea.

I cover winter crops inside my unheated polytunnel with lightweight row covers. Why the extra layer? In the unheated structure, the fabric offers another level of protection, further insulating the vegetables from the frigid winter temperatures. Instead of laying these covers directly on top of the polytunnel crops, which can damage the leaves and growing tips, I install wire hoops over the beds to support the fabric.

Cleaning Garden Covers

Regular cleaning and tidying of garden covers is vital to promoting healthy growth. Dirty or algae-covered polyethylene or polycarbonate blocks light, and debris can harbor pests or diseases. So give covers a good scrubbing whenever they get grimy!

If you've bought a greenhouse or polytunnel kit, read the manufacturer instructions before cleaning and follow recommendations for approved cleaning materials. Greenhouse supply stores carry cleaning products formulated for greenhouse materials, but I generally use plain water or a small amount of natural, mild dish soap.

Polyethylene. Polyethylene covers on permanent structures like domes and polytunnels need to be cleaned at least once or twice a year. Pick a drizzly, mild day where the cover has been "presoaked" by Mother Nature. This helps the dirt come off easier. Use a long-handled soft car wash brush or sponge to loosen dirt, rinsing well with clean water.

The trickiest bit is reaching the top of the tunnel. For this, invite a friend over and grab an old bedsheet. Tie a rope to each end of the sheet, tossing one of the ropes over the structure. Spray the top with water to loosen dirt and prewet the polyethylene. With someone on each side of the tunnel gently pull the sheet back and forth in a sawing motion, moving down the length of the tunnel. Rinse with water.

Polycarbonate. Wipe down with water or a small amount of mild dish soap mixed with water in a bucket. Use a soft cloth or sponge and rinse well.

Glass. Clean both the interior and exterior of glass structures with a sponge and soapy water. Reach grime or algae in cracks and crevices with a dishwashing brush, toothbrush, or another soft-bristle cleaning brush. Rinse well and avoid splashing soap on garden beds or plants. As you clean, inspect the glass for cracks, loose panels, or damage.

Keep it taut. Sagging, droopy, or wrinkling covers are more prone to flapping in the wind and rubbing against the frame. This can reduce the life expectancy of the film.

Use the right fastener. A fastening system like channel lock and Wiggle Wire causes less damage to film than using wood laths and screws.

Insulate with anti-hotspot tape. When installing polyethylene, apply anti-hotspot tape over metal hoops of walk-in structures. This foam tape is placed where the film comes in contact with steel hoops to prevent wear and tear.

Inspect it regularly. Get in the habit of giving the polyethylene a visual inspection every month or two, especially after a storm or bad weather. Check the film around door or window frames often as it experiences more friction and stress.

Repair tears right away. If you spot any tears or holes, fix them immediately with repair tape. I'd advise keeping a roll of repair tape on hand, so that you have it when you need it. A tiny hole can quickly tear into a large gash, particularly in high winds or during a storm. When repairing minor damage, apply the tape to both the inside and outside of the cover. Make sure the cover is completely dry before applying tape.

Look around. Ideally, this step should be done before you choose your structure site, but vegetation grows and if you haven't replaced your polyethylene cover in a few years, nearby trees or shrubs may have grown large enough to come in contact with the film.

Prevent snow buildup. If you live in a region with heavy snowfall, have a plan on removing it from your structure in a timely manner. I stand inside the tunnel using a soft broom to gently bounce the snow off the roof. I start in the middle, working my way to either end. You can also use a towel-wrapped rake (make sure all points are covered!) to bounce the snow from the inside or drag it off from the outside.

Keep an eye on the snow level around your structure. Snow can pile up around the perimeter of a dome, greenhouse, mini hoop tunnel, or polytunnel very quickly, making it more difficult for fresh snow to slide off the structure. This not only puts the film and structure in danger, but it also blocks valuable light. If you notice snow piling up, clear the area with a shovel or snowblower.

A dusty or dirty polytunnel or greenhouse cover blocks light. I give the polytunnel cover a thorough wipe down in spring and spot clean whenever I notice dirt during the growing season.

When placing repair tape on a poly cover, make sure the surface is dry and tape both the inside and outside surfaces.

Ash from a nearby bonfire burned a hole in the polyethylene of my polytunnel. Tears and rips are also common so it's a good idea to keep a roll of greenhouse repair tape on hand.

OTHER OPTIONS FOR LARGE STRUCTURES

Polytunnels are the workhorse of the garden, offering excellent cost savings and plenty of growing space, but their utilitarian design and large size may not be welcome in all neighborhoods. Greenhouses are generally more aesthetically pleasing than polytunnels, but they also cost a lot more to buy or build. While polytunnels and greenhouses are the most common large-size structures used in home gardens, they are by no means the only ones to choose from.

For those budget-minded gardeners who want to combine cost, practicality, and aesthetics there is the geodesic dome: an attractive structure with excellent strength and wind resistance. However, those wishing to take advantage of geothermal energy as well as solar energy may want to build a walipini, also known as a pit greenhouse. Walipinis provide a sustainable structure with fewer of the massive temperature swings of a greenhouse, polytunnel, or dome. Bioshelters are another type of larger-size garden structure popping up in home gardens, urban farms, and community gardens. The goal of a bioshelter is to create a self-sustaining ecosystem that may include animals like chickens, aquaculture, compost piles to enrich soil and provide heat, and heat sinks like rocks or water-filled barrels.

Bioshelters

A bioshelter is more than just a greenhouse; it's a self-sustaining ecosystem where plant and animal communities grow together. It may have habitats for chickens or rabbits, tanks for aquaculture, and space to grow long- and short-season crops like fruits and berries, as well as vegetables and herbs. You may even find a few compost bins or vermicomposting taking place inside a bioshelter.

The various communities inside a bioshelter are interconnected. Animals, like chickens and rabbits, produce manure to be composted, but they also provide body heat to warm the structure. The decomposing manure, and other compost materials, also give off heat. The water used for aquaculture acts as a heat sink, but so do rocks or barrels of water which may be included in a bioshelter design. At night, the stored heat is released back into the structure, raising the temperature. Over a 24-hour period, there are fewer large temperature swings in a bioshelter than there are in a traditional greenhouse or polytunnel.

Bioshelters are custom designed to suit their region, site, and purpose, but they do share a common feature (or lack of a feature) in that they are not fully covered in a transparent glazing material. The north side, and often the east and west sides, of a bioshelter are made out of a sloped

INSIDE THE BIOSHELTER

AQUACULTURE
600 Gallon Total Volume
1 Fish Production Tank
1 Vegetable Production Tank
1 Biofilter Cell

RAIN WATER HARVEST & THERMAL MASS
400 Gallon Total Volume
Harvest Water from Roof

CLIMATE BATTERY
Low Watt In-line Outflow Fan
2 Air Stacks to 2 Plenums
Gravel Heat Sink: 5 Cubic Yards

This bioshelter (also pictured on the facing page), designed and built by Regenerative Design Group in Massachusetts, incorporates raised beds, a 150-gallon aquaculture tank, a worm-compost trench, a rainwater collection tank, and a chicken coop/run.

hill, heat-absorbing rocks, or insulated walls. Sometimes the bioshelter is even dug straight down into the earth to insulate and reduce heat loss. The south side of a bioshelter, however, is covered in a transparent glazing material — typically double-paned glass or twin-wall poly-carbonate to reduce heat loss — and is angled toward the sun to capture as much solar energy as possible, particularly in winter. At night, the south side may be covered with thermal curtains to retain heat and reduce loss.

A typical greenhouse or polytunnel has no insulating walls, so a lot of heat is lost at night. A bioshelter, on the other hand, aims to strike a balance between gathering light and holding on to collected heat.

Geodesic Domes

A geodesic dome is a half-globe structure made of short struts arranged in repeating patterns of tri-angles. Triangles are often called the "superhero of shapes" because their fixed angles are naturally rigid. Personally, I think the visible triangular pattern of the struts combined with the half-globe shape make geodesic domes a beautiful addition to a backyard garden.

Domes aren't just pretty, they're also energy efficient, super strong, and well suited to growing food, especially in small spaces. The round shape maximizes light capture, allows wind to easily flow over and around its curves, and sheds snow easily.

Types of domes. As with polytunnels, there are various geodesic dome kits available in a range of sizes. There are also many DIY plans available online. The triangular struts can be made from wood or metal and covered with polyethylene or polycarbonate.

For a home garden, 15-, 18-, or 20-foot-diameter domes are common, but larger domes are available from greenhouse supply companies. Most domes are mounted on a low pony wall to raise the head-space. The middle height of a dome is half the diam-eter, so a 15-foot-diameter dome is just 7½ feet tall. Adding a 1- to 2-foot-tall foundation wall raises the height to accommodate medium to tall crops, which could otherwise butt up against the inside sidewall.

If you'd like to build your own dome from scratch, do your research first and find or buy a plan. There are a lot of precise measurements and cuts neces-sary for building a dome, and having a plan saves time, money, and frustration.

CAM & ANDREA'S DOME

There were several reasons why Cam and Andrea Farnell wanted to build a 26-foot-diameter dome in their backyard, but the deciding factors were their desires for a structure that was strong enough to stand up to heavy snow loads and strong winds, and insulated enough to retain heat through the winter. Their dome has done just that. In winters when other gardeners' polytunnels were collapsing from snow, their dome carried on. And on a sunny winter day when it's 5°F (–15°C), the inside temperature will be over 68°F (20°C). The soil never freezes and they've yet to see frozen ice on top of their 800-gallon water tank (which serves as both a source of irrigation during the winter and thermal mass to retain heat).

The dome structure is better at retaining heat than a regular greenhouse is, but the Farnells have also incorporated other heat-saving strategies. The dome was constructed on a cement base with knee walls, which act as a thermal mass to moderate temperatures. They've covered the north wall with reflective insulation. And the polycarbonate panels that came with the kit are 16 mil thick and five-wall construction: there are five layers of plastic with four air spaces in between.

Cam and Andrea have also tucked a rotary compost in the dome, which Cam calls a win-win. "The composting food gives off heat, which helps heat the dome," he says. "On warm days in fall and winter the heat in the dome keeps the composter going which it wouldn't if it was outside." And they use a small solar-powered fan to blow warm air from the top of the dome down into a pair of pipes that run under the raised beds along the outside wall of the dome, warming them.

The heat of summer can be a challenge; to bring temperatures down from the 90s (30 to 35ºC) to a more crop-friendly range, the Farnells use two thermostat-controlled box fans in the screen door, as well as four vents with automatic openers.

Overall, though, they're very happy with their dome. "We've been exceptionally pleased with how the dome works as a winter structure and have grown a lot of food in it," says Cam.

Walipinis

If I can ever find the space — and the time — I'd be tempted to construct a walipini. Sometimes called an earth-sheltered greenhouse, a pit greenhouse, or a solar pit, a walipini is an underground greenhouse 6 to 8 feet deep that uses solar energy and geothermal heat to create a protected environment. It is really just a hole in the ground with a clear top.

Why grow underground? There are fewer temperature swings and less heat loss than in a structure like my polytunnel. Thanks to their low cost and high effectiveness, they're finding a place in the home garden. Use them to extend the harvest by months in autumn, if not all winter long in most regions, as well as to get a jump on planting in early spring.

The design of a walipini depends on your location. In my northern garden, it would need to have a steep roof angle (around 60 degrees) to account for the low winter sun. Closer to the equator where the sun is higher in the sky, the roof would have a shallower angle.

These underground greenhouses can include passive heat storage (barrels of water, rocks), reflective curtains hung against the inside north wall, and insulating sheets of foam boards mounted on the underground walls.

When scouting a good location for a walipini, look for a site that has excellent drainage, especially in the spring when snowmelt could be a problem. Avoid areas where frost tends to pocket (at the bottom of a hill) and, if possible, pick a spot that is sheltered from strong winds.

JOE & MEGAN'S WALIPINI

The heart of Joe Hood and Megan Andrus's garden is a 10-by-16-foot semiunderground structure — a walipini. "It's designed not only to capture heat during the day but also to store some of that heat so it can release it at night and slow the cooling process," says Joe.

Joe and Megan's walipini is heated only by the sun and is designed to capture and retain heat for as long as possible. A cement wall encloses the bottom of the structure, with R-10 insulation going down 4 feet to the frost line. Joe also points out that the roof and siding are insulated to an R-value of 27.

The floor is composed of 18 inches of grapefruit-sized rocks with a layer of crushed stone on top to create thermal mass for heat storage or cooling. A 100-watt fan forces hot air at the top of the structure into a pipe moving the air through the bed of rocks and then back into the structure at the top of the raised beds. The rocks absorb the heat during

the day and slowly release it back into the structure at night.

The structure runs east-west with a triple-layered polycarbonate wall facing south. The three layers boost heat retention and let 72 percent of sunlight pass through. Ideally, the optimal pitch of the south wall of a walipini is dependant on latitude. Sackville, Nova Scotia, where the couple lives, is located about 45 degrees north latitude. Joe and Megan have installed their south wall at a 47-degree angle to capture maximum heating at the beginning of fall and end of winter, while still allowing 93 percent of available solar energy to enter the structure on the shortest day of the year.

Thanks to technology, they can keep a close eye on the growing conditions inside the structure around the clock; Joe monitors the temperature and humidity from his cellphone. "It's fascinating to discover how quickly it heats up, even on a cloudy day," he says.

4

Growing the
COVERED
GARDEN

Now that we've gotten a firm handle on all the different types of structures and covers available, it's time to learn more about gardening with them. Each winter as I make my annual garden plan, I factor covers into that plan. It helps me time seed starting, pinpoint planting dates, and come up with strategies for foiling garden pests and reducing disease. In this section I offer ideas for using beds and containers inside polytunnels and greenhouses; how to best use the space inside your structure; and the best mulches for prewarming soil, suppressing weeds, and improving the soil.

My polytunnel beds are 3 feet wide, which allows me to reach the entire bed from one side.

LAYING OUT THE GARDEN

Considering which covers you'll use when deciding how to grow your vegetables will make it easier to integrate the various types of covers into your garden and ensure that you've spaced beds or rows with room to accommodate the structures. While it's tempting to lay beds out with a tight spacing to increase your growing area, it's best to leave several feet between and around each bed to make planting, general garden chores, and season extension easier. For main pathways, I prefer 4 feet, ample room for working and wheelbarrow access. Smaller side beds can be spaced a bit tighter, at 2 to 3 feet apart. Depending on the application and the season, my raised beds may be topped with mini hoop tunnels covered in row cover, shade cloth, bird netting (for birds and deer), or polyethylene, and having ample room to erect the hoops and cover the structure makes set up quick and easy.

For me, planning a new growing space is great fun! There are so many ways to custom design your covered garden: beds can be raised or ground level, permanent or temporary. Bed height depends mainly on the soil quality, drainage, and your design preferences. Whether or not your beds are raised, they should be narrow enough that you can reach across without needing to walk or stand on the soil.

The width of the polytunnel is a determining factor in the width and length of the interior beds. Small 8- or 14-foot-wide structures have space for two lengths of beds, or a single large U-shaped bed along the sides and back of the tunnel. Sixteen-foot-wide tunnels can accommodate three or more lengths of beds with two or more pathways between the beds.

Don't skimp on the pathways! It's tempting to make the paths narrow so that most of the interior space is for growing vegetables, but paths should be wide enough to allow easy access for the gardener. Paths that can accommodate a wheelbarrow make moving garden debris, soil, compost, and other materials fast and easy.

Beds aren't the only options. Many home growers opt to use pots or fabric grow bags, which can be moved around to change the layout from year to year. Remember, though, that pots dry out quicker than beds, so you'll need to be vigilant about watering.

In our tunnel, we went with hemlock-edged raised beds to lift up the soil level. Our existing soil was poor and very weedy so we excavated a foot down and removed it. We lined the trenches with professional weed barrier fabric and filled the beds with a mixture of garden soil, compost, chopped leaves, and composted seaweed.

SETTING UP RAISED BEDS

Rectangular raised beds are perfect for the protected garden. You can easily cover them with mini hoop tunnels topped with row cover, polyethylene, or shade cloth. You can even build a removable cold frame top to fit over a raised bed for spring, autumn, and winter harvesting. Raised beds in a walk-in cover like a polytunnel, dome, or greenhouse keep soil in place and pathways tidy.

There are lots of other reasons to grow in raised beds, too! Raised beds thaw quicker in the spring — and that can be sped up further by putting up a mini hoop tunnel or laying plastic mulch over the soil. Raised beds allow you to apply compost, aged manure, and other amendments more efficiently so that nutrients go directly to the vegetables. They offer improved

Raised beds offer many advantages to a vegetable gardener. They warm up earlier in spring, offer loose, non-compacted soil, and are the perfect size for garden covers like mini hoop tunnels.

Our RAISED BED Renovation

In the spring of 2016, we renovated and expanded our vegetable garden. We increased our growing space by around 30 percent and built twenty raised beds. Going from eight low, free-formed beds to twenty tall, wood-sided raised beds actually reduced my workload. Raised beds generally reduce weed growth (as long as you never let weeds go to seed) and my beds freed me from years of battling tenacious perennial weeds. And the older I get, the more my knees appreciate my raised beds; I don't have to kneel to seed, plant, weed, or cultivate.

Raised beds also make season extension a snap! Our 4-by-8- and 4-by-10-foot beds are the perfect size for erecting mini hoop tunnels for spring, summer, autumn, and winter protection. I use PVC or metal conduit for the hoops, covering them with the appropriate material, depending on the season and crop being protected. In spring and autumn, it's typically row covers, but in late spring and summer, I use shade cloth to extend the harvest of cool-season greens. Come winter, the tunnels are topped with greenhouse polyethylene and shelter crops like leeks, kale, mustards, winter lettuce, spinach, and scallions.

drainage, even after a downpour. They dramatically reduce soil compaction by keeping foot traffic to designated pathways. And they allow for more intensive planting, which results in higher yields and fewer weeds.

Use what you've got. Raised beds can be free-formed or edged with untreated lumber, bricks, cinder blocks, logs, rocks, or whatever other materials you have handy. Our beds are made from local, untreated hemlock and measure 4 by 8 or 4 by 10 feet. (I'd suggest sticking to a width no greater than 4 or 5 feet, so you can easily reach the middle without straining.) Most raised beds range from 6 to 12 inches tall, but we built ours 16 inches tall, partly because the existing soil was poor and the site was plagued with perennial weeds. Depending on your mobility, you can build beds as high as you like to allow easy access.

When erecting mini hoop tunnels over our raised beds, I generally insert the PVC or metal hoops directly into the soil in the beds. The support of the wooden bed edge helps keep the hoops in place. In spring and fall, often our season extenders are temporary; in place for a few hours, days, or weeks. It's quick and easy to just insert the hoops into the soil. In winter, the tunnels are more long-term, often sheltering vegetables for months, and they also need to be able to stand up to a snow load. For these reasons, I attach winter hoops to the beds with metal brackets.

A Word about Containers

Containers also deserve a place in the protected garden. For example, I combine both raised beds and containers in my polytunnel. The raised beds are permanent and run down each side of the tunnel while the middle section is reserved for spring, summer, and autumn containers. I like the versatility of containers, which allow me to adjust the planting plan and design of my polytunnel as needed. Trusses are conveniently located just above the containers and can support tomatoes, cucumbers, and other tall or vining crops. However, it's not just about using containers in walk-in structures. Pots, window

I use fabric raised beds to grow vining and tall vegetables down the middle of our polytunnel. They allow me to adjust the design of the growing space from year to year, and they hold a high volume of soil for vigorous tomato, cucumber, and melon plants.

boxes, and fabric planters can be put to work on your deck and patio and fitted with seasonal covers to allow earlier planting or later harvesting. Suspend a shade cloth with grommets and sisal rope above for summer shading or add a tomato cage covered in polyethylene to protect from frost in spring or fall.

Growing in containers offers some advantages to gardeners with less-than-ideal site conditions or unpredictable weather. They enable you to grow crops where they'll perform best, whether it's planting leafy greens on a patio in partial shade or peppers in a warm microclimate (I have a sunny backyard, but my back deck is a hot spot sheltered by a slope that captures heat, so I grow eggplants and peppers in planters there). Many containers can also be moved indoors quickly to avoid an onslaught of hail or an unexpected frost.

TEST YOUR SOIL

There are many reasons why regular soil testing — starting with when you first start a new garden — is a good idea. It gives you a snapshot into the health of your soil and helps you keep ahead of issues like nutrient depletion. It also helps guide the types and amount of amendments and nutrients you need to add to your beds. Even if my crops are growing well, I generally test my soil every 2 to 3 years.

Soil testing also helps you ensure that your soil pH is in the right range for growing vegetables. A pH of 7.0 is neutral, and 6.5 is the ideal for most food plants. When the soil pH falls below 5.5 or is above 7.0, nutrients and micronutrients are less available to plants. In my region, the soil tends to be acidic, so I raise the pH with an application of limestone. Gardeners in regions with alkaline soil can lower their soil pH with elemental sulfur.

SPRING PLANTING IN THE COVERED GARDEN

Protective structures give you some wiggle room when it comes to planting seeds or seedlings in the garden. Covers can push back the spring planting season by weeks (row covers, cloches, mini hoop tunnels) to months (polytunnels, greenhouses, domes, cold frames) and extend the harvest into late autumn or winter by the same time frame.

I find it's easiest to divide the year into three growing seasons: warm season, cool season, and cold season. The span of time between the last spring frost and the first autumn frost is the warm season, spring and autumn are the cool seasons, and winter is the cold season.

Simple plastic or glass cloches are handy for covering heat-sensitive pepper or tomato seedlings in spring.

Exactly how much earlier you can plant beneath a cover will depend on the type of cover and its materials. Twin-wall polycarbonate, which tops my cold frames, is more insulating than a single layer of polyethylene, for example. I seed spinach in a cold frame in February, while my spring garden spinach isn't seeded until late April, giving me a 2-month head start. That early planting also applies to my polytunnel. Just as we're finishing up harvesting winter crops in February, new seeds for carrots, beets, Asian greens, and peas are sown in the open spaces left behind.

But it's not always about the cold-season veggies — garden covers allow you to plant warm-season crops like tomatoes, peppers, and cucumbers weeks earlier, too. My tomato seedlings are set out in my unheated polytunnel in early May (about a month earlier than unprotected garden tomatoes) but could be planted even earlier if I used a small heater for the night-time temperature dip. Without additional heat, I keep row covers handy for those May nights when the temperature drops below 50°F (10°C).

Days to maturity. To figure out timing, make a list of which crops you want to grow and how long they take to mature. Each crop has a specific time frame it needs to go from seed or transplant to harvest; this information is listed as "days to maturity" in seed catalogs.

For example, 'Provider' bush bean takes 50 days to go from seed to harvest. It is a warm-season vegetable and can't be seeded until after the last spring frost. 'Sungold' tomato takes 57 days to maturity, but that time frame is from the transplanting date, not the seeding date. Read seed catalogs carefully to be sure you've got the right planting information.

Choose the season. The next step is to decide when you want to grow the crops. Warm-season crops need to be grown between the last spring

It's quick and easy to build a mini hoop tunnel over tomato beds in spring. Whether covered with a row cover or a polyethylene sheet, the tunnel creates a microclimate and protects seedlings from the variable temperatures of May.

frost and the first fall frost. My growing season (the days between the frost dates) is approximately 140 days so I need to choose crops that mature in 140 days or less.

Using Covers to Get a Jump Start on Spring

Warming soil with plastic sheeting. There are plenty of vegetables that can be direct seeded in garden beds once the soil temperature is above 45°F (7°C). The list includes carrots, beets, arugula, spinach, peas, and radishes. Yet spring soil, especially soil that is saturated with cold or icy water, takes longer to warm up. A savvy gardener can help speed up the warming-up process with a plastic sheet laid directly on the bed.

There are several ways to prewarm the soil. The easiest is just to cover the soil with a clear or black plastic sheet for 12 to 14 days before you intend to plant. Lay the sheet on the soil surface and weigh it down with rocks or pin it with garden staples to keep it secure on windy days. Leave it on day and night until you're ready to plant. Once you remove the plastic, you'll likely notice some weed growth. That's okay! Pulling those weeds now means fewer weeds in the future.

Use plastic sheets to prewarm soil before planting out warm-weather crops like melons, eggplants, and peppers. Setting these seedlings in the garden while the soil is still cold can damage or even kill the plants and prewarming is a simple way to boost success, especially in northern climates like mine. Once the plastic sheet is removed and the bed seeded or planted, you can follow up by erecting a quick mini hoop tunnel over the bed as extra insurance against the uncertain spring weather.

Starting seeds outside, under cover. Whether you prewarm your soil for seeding or just allow it to warm up on its own, you can still use a variety of covers to start seeds earlier. I generally use mini hoop tunnels, row covers, and cloches to speed up my spring garden seeding. Salad greens like lettuce and spinach, which are normally direct seeded in the open garden in late May in my region, can be sown in beds 8 to 10 weeks earlier with a mini hoop tunnel covered in polyethylene. Row cover is less insulating but can be used to cover hoops or laid directly on beds to permit planting 4 to 6 weeks earlier. Vent mini tunnels or remove fabric covers during the day when temperatures are above 45°F (7°C).

Cold frames can be used to grow crops, but they can also be used to start seedlings in early spring, or harden off seedlings before they're transplanted to the open garden.

Cloches can also shelter seeds, but work best in containers or when small clumps or seeds are planted. With their smaller size and often round or square shapes, they're harder to use when seeding rows or bands in garden beds. That said, cloches are still super handy in spring, but I use them more for protecting individual seedlings.

Using Covers to Help Harden Off Seedlings

Seedlings started indoors under grow lights have a pretty good life: no excessive heat or cold, ample food and water, and no insects or deer munching on their leaves. As a result, they haven't had to develop their horticultural suit of armor: the waxy leaf cuticle layer that protects them from sun and wind. "Hardening off" is the process of gradually introducing these pampered seedlings to outdoor growing conditions, so that they have time to build up their armor. It should take 5 to 7 days.

What if you're planting seedlings under a garden cover like a polytunnel or cold frame, and not the open garden? Do you still need to harden them off? The short answer is "yes." If I planted my indoor-grown seedlings directly into the polytunnel beds, they would quickly show signs of shock; wilting, bleached leaves, or curled foliage.

Garden covers can help plants adjust during the hardening-off process. Sometimes real life can come as a shock to plants that have been pampered indoors under a grow light. There are the up-and-down temperatures, gusty winds, downpours, hail, and munching insects and mammals to contend with. Use garden covers like mini hoop tunnels, row covers, shade cloth, cold frames or even a polytunnel to make the transition less stressful on your plants. They'll provide shelter from inclement weather and pests while the seedlings adjust to life in the garden.

A Schedule for Hardening Off Seedlings

DAY 1: Place seedlings outdoors on a mild day in a fully shaded spot about a week before you intend to transplant them. Bring them indoors for the night. If you've got plenty of seedlings to harden off, place a shade cloth–covered mini hoop tunnel in a corner of your yard to serve as a hardening-off zone.

DAY 2: Give seedlings full to dappled shade. Bring them indoors for the night.

DAY 3: Place the seedlings in an area where they are exposed to several hours of morning sun. Leave them out if the temperatures are mild. You can also cover them with a row cover for the night for extra protection. Or, if you have a shade cloth hardening-off zone, cover the entire structure with polyethylene to protect seedlings at night.

DAY 4: Allow seedlings a half day of full sun. Again, leave them out overnight if the temperature is mild.

DAY 5: Give seedlings a full day of sun. Leave them out overnight.

DAY 6: Transplant them in the garden or under a garden cover like a mini hoop tunnel, cold frame, or polytunnel.

If you have no choice but to transplant seedlings into the garden before they're fully hardened off, float a piece of shade cloth or row cover above the bed to minimize shock.

USING COVERS
as Insurance

After plants are hardened off and in the garden, think of covers as your insurance; they can be called to action if the weather turns. I can't tell you how many times I've tucked my tomato seedlings into the open garden after the last expected frost, only to learn that frost is in the forecast! Row covers and cloches help protect from frost and cold weather (spring, autumn), shade cloth protects from heat (summer), and cold frames, mini hoop tunnels, and polytunnels protect from heavy rain, frost, and snow (spring, autumn, winter).

GROWING VERTICALLY IN LARGER STRUCTURES

Growing vertically in walk-in garden covers like polytunnels, greenhouses, and domes is a smart way to maximize space and grow more food. But there are other benefits, too, such as a decrease in disease and insect infestations.

Three Benefits of Vertical Growing

Maximize space. Space is at a premium in polytunnels and greenhouses, so it makes sense to use all that empty vertical space for tall or vining vegetables.

Grow more food. Whether growing in the open garden or under cover, vining vegetables can be planted closer together when they're trained to grow vertically. It's also easier to spot the fruits, and those like cucumbers, snake gourds, or 'Tromboncino' summer squash will have straighter fruits when they're grown vertically.

Fewer pests and diseases. Trellising vegetables like tomatoes and cucumbers can reduce the occurrence of pests and diseases. It keeps the plants from flopping on the ground (which also means a tidier space that is easier to work in), increases air flow around the foliage (which

The steel trusses at the top of my polytunnel add strength to the structure but also provide a support for my tomatoes and cucumbers. Wires run across the trusses and I hang strings from the wires. As the vines grow, I tie or clip them to the strings.

speeds up drying from condensation or water splashed during irrigation), and makes it harder for insect pests and slugs to damage the plants.

Planning for Vertical Crops

It's a good idea to think about vertical crops early in the planning process — as in, before you even buy or build a structure.

Consider crop bars. Many polytunnel and greenhouse kits come with crop bars or have them as an option. Crop bars are horizontal struts that run across each arch above head height. They add strength to the structure and support vining vegetables grown up trellises, string, twine, or wires. You don't need crop bars to grow vertical vegetables, but they are handy — and strong. In domes or DIY structures without crop bars, use trellises or stakes to support tomatoes and cucumbers.

Plant strategically to avoid shade. When planting seedlings inside a structure, it's important to keep mature size in mind so that they can be spaced appropriately and planted in a spot where they won't shade other vegetables. It's not just tall, vining crops that can cast shade; vertical elements like shelving, hanging baskets, and trellises do, too. In most structures, some shading is inevitable, so tuck shade-tolerant vegetables in areas with less light. Edibles that can be grown in partial shade include salad greens; scallions; root crops like beets, radishes, and turnips; and herbs like chives, cilantro, and parsley.

Train Plants as They Grow

They may start off small, but tomato, cucumber, and melon seedlings soon grow tall and heavy, especially when laden with fruits. Install your support system before plants are in the ground and train the plants to the structure as they grow. Just remember to match the crop to the support method. Climbing vines like pole beans, peas, cucamelons, and squash quickly scramble up trellises, netting, and tripods. Vining tomatoes need help to stay upright; they can be staked or trained to grow up strings or twine.

Use untreated twine to support vining vegetables. At the end of the growing season, they can be tossed on the compost pile with the plant debris.

THE WIRE-AND-STRING METHOD

This method utilizes wire run along the crop bars in a polytunnel, with plants trained to individual strings; I use 9-gauge galvanized wire, secured before planting, and untreated sisal twine, which can be composted at the end of the season. Buy good-quality garden twine, not an inexpensive product from a dollar-type store. Why? I've found cheap twine to be more likely to break once the plants are large and heavy with fruit. Walking into your tunnel, greenhouse, or dome and seeing a mature tomato plant laying on the ground is devastating, but preventable if you choose high-quality twine.

Once seedlings are in place, hang lengths of twine above each plant. Tie these loosely to the bottom of each plant or use a clip to fasten them

	Trellis	Netting	String	Cages	Stakes
Cucumbers	X	X	X	Dwarf only	
Indeterminate tomatoes	X		X		X
Cucamelons and burr gherkins	X	X			
Melons	X	X	X		
Peas	X	X			
Pole beans	X	X	X		
Eggplant			X	X	X
Peppers			X	X	X
Edible gourds	X	X			

to the stem. As the plant grows, wind the stem around the string. Always wrap the stem in the same direction so the plants don't unwind and fall down.

ADDITIONAL SUPPORTIVE STRUCTURES

The wire and string method is popular, but there are many ways to support vertical vegetables in walk-in structures:

Wire or wooden cages. I rarely find traditional wire tomato cages tall or sturdy enough for indeterminate tomatoes, but they're handy for keeping pepper, eggplant, and bush cucumber plants off the ground. Taller metal or wooden cages can be made or bought for larger plants.

WOODEN STAKES

Stakes. In my raised-bed vegetable garden, I use 7-foot-tall wooden stakes to train and support indeterminate tomatoes like 'Sungold'. They're inexpensive and can be used over several seasons. You can also use metal stakes, rebar posts, or spiral tomato stakes available through garden supply stores.

Trellises, tripods, and obelisks. There are many different types, sizes, and styles of trellises, tripods, and obelisks available. I've used wooden and metal obelisks, bamboo post tripods, and one- and two-panel trellises.

Pea and bean netting. I like to use the north end of my polytunnel to grow pole beans, peas, and cucamelons on nylon netting. I don't have beds at the end of my tunnel, so I plant them in large containers or fabric planters.

OTHER WAYS TO GROW VERTICALLY

Shelving. Whether freestanding or hung from crop bars or gable ends, greenhouse shelves can hold pots of strawberries, herbs, or vegetables. They're also a handy place to store tools and supplies or start trays of seedlings.

Living walls. Living wall systems are perfect for compact edibles like strawberries, herbs, salad greens and trailing tomatoes like 'Lizzano'.

Many systems contain a water reservoir, which makes it easier to keep plants hydrated in a greenhouse or polytunnel. Create your own living wall with wall planters or window boxes hung on wooden gable ends.

Vertical containers. You can make tall vertical containers or towers from a variety of materials or purchase them from garden supply stores. Many have pockets all around the outside to hold dwarf or compact crops.

Hanging baskets. Crop bars are a convenient spot for hanging baskets. Baskets dry out quickly, though, particularly in warm weather, so be sure to irrigate regularly.

Pruning Vertically Grown Plants

Closely planted vertical crops will need to be pruned or pinched regularly. Pruning removes parts of the plant like suckers, buds, or shoots to encourage healthy growth and increase flowering and fruiting. It permits higher-density planting; you can fit more plants into your greenhouse or polytunnel if they're pruned well. And it increases air circulation around plants, which helps foliage dry quicker and reduces the risk of disease.

Pruning isn't a one-time deal, however; it needs to be done on a regular basis. The most important pruning task is to remove suckers from tomatoes. Suckers form at each leaf node, and if each one was allowed to grow, the plant would quickly become a mass of foliage, flowers, and fruits. This delays harvest and reduces fruit size.

My favorite pruning tool is my fingers. When suckers are small, it's easy to remove them with a quick pinch. If you neglect pruning for a week or two, you may need pruners to remove the larger suckers. I also prune out the lower leaves of tomatoes as they yellow. This increases air circulation and reduces disease.

For crop-specific pruning information, turn to Part 2, which begins on page 140.

Supporting LARGER FRUITS

Small and medium-size fruits like tomatoes, cucumbers, eggplants, and peppers don't need supplemental support when grown vertically, but the heavy fruits of melons, squash, and gourds benefit from a simple sling. As fruits develop, tie lengths of pantyhose, cut up T-shirts, old row covers, or other stretchy fabrics to the trellis to prevent breakage or damage.

MULCHING

Mulching the soil is a technique that can increase yields, cut down on watering and weeding, add organic matter to the soil, and minimize disease or insect infestations. Most food gardeners use a biodegradable mulch like straw, grass clippings, shredded leaves, or compost. There are also plastic mulch options, although these are more common in commercial growing operations.

The key to getting the most from your mulch is applying it at the right time and at the right depth. Organic mulches cool the soil and mulching too early in the season can prevent the soil from warming up, especially in northern gardens. Most types of plastic mulch, however, speed up soil warming, which can be useful for those same northern gardeners.

Using a straw mulch in my polytunnel is the best way for me to reduce watering. Bare soil dries out very fast when the late spring and summer tunnel heats up to 80 to 95°F (30 to 35°C). A 2- to 3-inch layer of straw holds the soil moisture and has cut my watering in half. When applying an organic mulch, wait until the soil has warmed in mid to late spring. Because I want my mulch to hold soil moisture, I wait to apply it until we've had a nice, deep rain; or, if mulching in a greenhouse or polytunnel, after a deep watering. Although mulch discourages weed growth, you should never apply it on top of weeds. You need to weed first.

Although mulch is an asset in the summer polytunnel, I remove it when the cooler temperatures of autumn arrive. I find this is the time when the slug population makes a comeback and the straw provides a perfect place for the slugs to shelter.

You can also pair mulch with a structure like a mini hoop tunnel covered with polyethylene, row cover, insect barrier, or bird netting. The mulch holds moisture and reduces weeds, while the cover protects from frost, insects, rabbits, or deer, depending on the cover.

Apply the mulch around plants, but not directly against the stem or leaves. For organic materials like straw or shredded leaves or newspapers, add a 3- to 4-inch layer of mulch. For a mulch of compost or grass clippings, apply 1 to 2 inches to the soil surface. If slugs are an issue in your garden, try mulching with plastic film, which is less hospitable to slugs than loose organic material.

COMPARING MULCHES

MULCH MATERIAL	PROS	CONS	NOTES
Shredded leaves	► Easily sourced ► Inexpensive ► Adds organic matter as it breaks down ► Inviting to worms	► Some leaves, like oak, decompose slowly ► Black walnut leaves contain juglone, a growth-inhibiting chemical	► Gather leaves in autumn, shredding and bagging them for spring mulch ► Use leaves to create leaf mold compost, which also makes a fantastic mulch ► Avoid leaves gathered from a lawn where dogs are allowed to poop
Straw	► Easily sourced ► Inexpensive ► Adds organic matter as it breaks down ► Deters pests that lay their eggs near the soil (flea beetles, squash vine borers) ► Lasts an entire season ► Keeps fruits like cucumbers and melons clean	► Can be a challenge to source spray-free bales ► Can have weeds from fields or seeds that sprout ► Can provide shelter for slugs in early spring or autumn when inside cover temperatures are cool	► Only use spray-free, seedless straw as a mulch
Grass clippings	► Easily sourced ► Inexpensive ► Adds organic matter as it breaks down	► Decomposes very quickly ► Can contain lawn weeds	► Avoid grass clippings from lawns that have been treated with pesticides
Plastic sheeting	► Warms the soil (can be applied earlier than organic mulches) ► Effective weed barrier ► Holds soil moisture	► Doesn't allow water to pass through, need to use soaker hose beneath mulch ► Can overheat soil in warm climates ► Unattractive appearance ► Biodegradable films can leave plastic bits in the soil	► Secure with staples or weigh down with rocks or logs
Fabrics	► Can be laid between rows or beds to suppress weed growth ► Water permeable	► Expensive (professional grade) ► Repels earthworms ► Unattractive	
Newspaper (shredded)	► Easily sourced ► Inexpensive ► Reduce, reuse, recycle!	► The ink on glossy, colored pages can contain heavy metals. Use only black and white, uncoated sheets.	
Compost	► Adds organic matter as it breaks down ► Easy to make ► Quick to break down	► Can be difficult to make enough compost for mulching ► Takes several months to make compost	

USING PLASTIC MULCHES TO IMPROVE THE HARVEST

You've probably noticed the rainbow of colored plastic mulches available in garden supply stores and catalogs. But are these red, silver, green, or white plastic films effective in a food garden? It depends on your goals. I've found organic mulches like straw or shredded leaves to be just as effective as plastic mulch for blocking weed growth, and I think natural materials offer a more appealing aesthetic to my garden.

That said, experimenting with a red plastic mulch beneath your tomatoes or green for melons, especially in a short-season region, can result in increased success. Plastic mulches can warm soil, reduce weed growth, speed up ripening, or increase yield, depending on the product.

Professional growers often use plastic mulch in polytunnels to promote a heavy yield and maximize crop performance. However, plastic mulch, and black plastic mulch in particular, can heat up too much in spring and summer, stressing out your plants. To avoid this, greenhouse growers often use white or white-on-black plastic mulch inside structures. The white keeps the soil cool and the black prevents weed growth. Plastic mulch also makes watering more difficult, as it requires drip irrigation or soaker hoses beneath the plastic film.

Black. The most common type of plastic mulch, black is readily available at most garden centers or garden suppliers. It warms soil by several degrees and is excellent at suppressing weed growth. Best in northern areas. Because it is a petroleum product, it's difficult to recycle.

White on black. This is often used in professional greenhouse operations for its ability to suppress weeds and keep the soil cool. It's also popular with southern gardeners as it doesn't heat up as much as black film. The white side reflects heat away while the black on the bottom prevents weeds from growing.

Red. Generally used as a mulch around tomato plants, red mulch can increase yield by up to 20 percent. It reflects far-red wavelengths up to the foliage, which initiates photosynthesis and encourages growth.

Silver. Silver mulch is often used in pepper and eggplant beds to repel common pests like aphids, flea beetles, and thrips. It can also increase yield by up to 20 percent, according to trials at Penn State. The bottom side of the film is black; the combination of silver and black cools the soil by several degrees, making this a good mulch for warm climates.

Clear. I've had good success using clear plastic to warm the soil of my tomato, pepper, melon, and eggplant beds in spring. In fact, it's more effective than black plastic at soil warming, but clear plastic doesn't discourage weed growth and it can become a weedy mess under the film.

Green. Green plastic film is another crop-specific mulch, but this one works best with heat-loving squash, melons, and cucumbers. University trials have shown it to encourage earlier ripening and boost overall yield. It's effective at soil warming and blocking weed growth.

Red plastic mulch can help increase the productivity of tomato plants.

PLANTING FOR WINTER

In order to plan for winter harvesting, you must have a basic understanding of how day length affects the rate of growth in plants. Every day of the year, day length changes, with days becoming shorter and darker in autumn and longer and brighter again in late winter. Day length is also determined by your latitude. The farther north you are, the more extreme the shift; so, places like Sweden or northern Alaska have very long days in summer and very short days in winter. Knowing your latitude allows you to calculate the day length for any day of the year.

Once day length dips below 10 hours, the growth of most plants slows considerably. This isn't as important for spring and summer crops, but if you're interested in harvesting during the fall and winter, you'll need to know the date when your area will have less than 10 hours of sunlight a day. Ideally, fall and winter crops should be planted so that they are about 90 percent mature by the time the day length falls to less than 10 hours. They can then be harvested from their season-extending covers for weeks, months, or all winter long depending on your region and the crop.

At my Halifax, Nova Scotia, location (44° 52' latitude), my day length falls below 10 hours on November 5 and does not go above 10 hours until February 6. I therefore aim to have my crops reach 90 percent maturity by November 5. Calculate your own day length with an online sunrise and sunset calculator.

How to Determine Planting Dates for Winter Harvest

You may be wondering how you can calculate the proper planting date. It's actually quite easy — no math degree required! You need to know three pieces of information: 1) the length of your growing season; 2) the average first autumn frost date; and 3) the days to maturity for your desired crop.

Winter is the sweetest season for carrots! I deep-mulch my carrot beds with straw so we can harvest varieties like 'Napoli' all winter long.

Let's use this information to figure out when to seed sweet 'Napoli' carrots, my family's favorite fall and winter carrot. Then you can use this trick to determine the planting dates of cool- and cold-season vegetables for your fall and winter gardens and garden structures.

PLANNING A CROP OF 'NAPOLI' CARROTS FOR FALL

► Length of growing season: 140 days
► Average first frost date: October 10
► Days to maturity for 'Napoli' carrot: 58 days

Based on its days to maturity, we know that our 140-day growing season can accommodate 'Napoli' carrot. But because the growth of most vegetables slows down in autumn with the shortening days, I add an extra week to the listed "days to maturity" to account for this slowdown. So, I now expect our 'Napoli' carrot to take 65 days to go from seed to harvest.

If you count backwards 65 days from our first expected frost date of October 10, you land on August 6. This is when I need to seed my fall/winter crop of 'Napoli' carrots.

Niki's Under Cover CALENDAR

January

► Plan out bed and polytunnel planting schemes, taking crop rotation into account

► Order seeds

► Harvest greens, root crops, scallions, and other cold-season vegetables and herbs from the winter covers and mulched beds

February

► Harvest

► Putter and tidy around the polytunnel when the temperature allows

► Inspect polytunnel structure and polyethylene for winter damage, holes, loose screws, etc.

► Sow seeds under grow lights for artichokes

► Prep any empty beds in the polytunnel for end-of-the-month seeding

► In late February, sow seeds in amended polytunnel beds for cold-season greens (kale, arugula, spinach, Asian greens)

March

► Ventilate cold frames, mini hoop tunnels, polytunnel, and other garden covers when the temperature climbs over 45°F (7°C)

► Sow seeds for cool- and cold-season vegetables in polytunnel and cold frames as winter beds are harvested and amended

► Plant early potatoes in polytunnel

► Sow seeds for most vegetables, flowers, and herbs indoors under grow lights

► Water new plantings as needed in garden structures

► As the weather warms, remove secondary coverings inside the polytunnel

► Watch for slugs in the polytunnel, handpicking as necessary

April

- ► Give the polytunnel floor a good sweep and scrub and set up containers or fabric planters down the middle of the tunnel
- ► Sow seeds for carrots, beets, greens, turnips, radishes, and scallions

- ► Plant seedlings of broccoli, cabbage, kale, and other cabbage cousins in garden beds, covering the beds with polyethylene-covered mini hoop tunnels

- ► Ventilate structures, water as needed
- ► Harvest the last of the over-wintered parsnips

May

- ► In early May, switch out the polyethylene mini hoop tunnel covers over the garden vegetables for row covers
- ► Switch out polyethylene covers for insect barrier over cabbage family crops to prevent cabbage moths from laying eggs
- ► Plant tomatoes, peppers, cucumbers, and other warm-season vegetables in polytunnel

- ► Keep row covers handy for nighttime protection
- ► Ventilate structures, water as needed
- ► Direct sow seed for root vegetables and greens in the garden, covering soil with a row cover to encourage germination (remove cover once seedlings emerge)
- ► Use shade cloth to harden off seedlings before they're moved to the garden

- ► Mulch soil with straw in polytunnel
- ► Cover garden potato bed with insect barrier to deter Colorado potato beetles
- ► In mid to late May, remove mini hoop tunnels from garden beds, but keep row covers handy, just in case the temperature dips again

June

- ► Start more seeds indoors under grow lights for succession planting and fall crops
- ► Keep row covers handy for unexpected temperature drops
- ► Once all crops are harvested from cold frames, sow a cover crop like buckwheat to build up the soil

- ► Fertilize garden biweekly with a liquid organic fertilizer
- ► Run strings from the trusses in the polytunnel for tomatoes, cucumbers, and other vining crops
- ► Stay on top of pinching out suckers on polytunnel tomato plants

- ► Remove lower leaves from tomato plants as they grow to deter early blight
- ► Erect a shade cloth–covered mini hoop tunnel over garden lettuces to delay bolting
- ► Keep an eye out for pest and disease problems

July

- Remove insect barrier over cabbage family plants as cabbage moth season has passed

- Continue to train greenhouse tomatoes, cucumbers, peppers, and eggplant on their strings or trellises

- Keep an eye out for powdery mildew on cucumbers and squash, particularly in the polytunnel (spray a milk-water solution weekly to reduce the risk of powdery mildew)

- Harvest polytunnel cucumbers and tomatoes as they ripen

- Harvest early potatoes from polytunnel

- Plant more bush cucumbers, zucchini, and bush beans in polytunnel for a September harvest

- Water potted tomatoes consistently to reduce blossom end rot

- Keep an eye out for pest and disease problems

- As fall crops are transplanted into the garden, cover the bed with shade cloth for a few days if the weather is hot and dry to reduce shock

- Hand-pollinate cucumbers, squash, and melons in the polytunnel as necessary (don't pollinate parthenocarpic cucumbers)

- Sow seeds for fall and winter carrots in cold frames, beds, and the polytunnel

- In late July, move cabbage cousin plants into the garden for fall and winter harvesting

August

- Water and fertilize regularly

- Sow seeds for beets and daikon radishes in the cold frames, beds, and polytunnel

- Harvest often

- Keep an eye out for pest and disease problems

- Continue to hand-pollinate crops in the polytunnel as needed

- Continue to remove lower leaves of tomato and cucumber plants in polytunnel and garden

- Keep clipping plants to their strings or trellises

- Sow seeds indoors under grow lights for greens like lettuce, chard, and kale

- Harden off seedlings under a piece of shade cloth

- Inspect and clean fabric and plastic garden covers to ready them for fall and winter use

September

- Transplant seedlings grown under grow lights into cold frames, polytunnel, and beds that will be covered with mini hoop tunnels
- Water and fertilize regularly
- Top tomato plants in the garden and polytunnel to encourage existing fruits and flowers to ripen before frost
- Direct sow seed for salad greens, spring radishes, turnips and other quick-growing vegetables in garden beds, cold frames, and polytunnel
- Dig up any herbs or vegetables you wish to move to the polytunnel for winter harvesting (parsley, kale, thyme, etc.)

October

- Continue to harvest polytunnel vegetables
- Close cold frame tops and polytunnel sides when night temperatures are below 55°F (13°C)
- Remove any spent crops in the garden beds, frames, and tunnel
- Water consistently but stop fertilizing
- In late October, sow seeds for overwintered spinach in cold frames, mini hoop covered beds, and the polytunnel to be harvested the following March

November

- Clean up tomato and cucumber plants that grew in containers down the middle of the polytunnel
- Empty pots, clean, and store for winter
- Remove twine strings from the polytunnel that were used for vertical crops
- Check structures before winter arrives to make sure they're in top shape
- Continue to water cold frame, mini hoop tunnel, and polytunnel vegetables
- If you have an automated watering system, drain and winterize it
- Set up wire hoops inside the polytunnel to act as a secondary winter cover for greens and other vegetables
- In late November, deep mulch garden root and stem vegetables like carrots, beets, celeriac, and leeks

December

- Enjoy the protected space of the polytunnel on mild days
- Harvest vegetables as needed
- Likely, the last watering of the year will take place in early to mid-December (once the ground freezes, crops will need little water)
- Cover wire hoops in the tunnel with a sheet of polyethylene or row cover; vent on mild days

MULCHING OVERWINTERED CROPS

Mulching root and stem crops for late autumn and winter harvesting is one of the easiest ways to extend the season. Winter mulch should be applied before the ground freezes, so keep an eye on the temperatures in mid to late autumn. It's harder to harvest if the beds are mulched after freezing.

Straw or shredded leaves. These are my mainstays. Mulch beds with at least a foot of straw or shredded leaves, then cover the mulched bed with an old row cover, bedsheet, or another fabric to hold the material in place. Secure the fabric using garden staples or rocks, logs, or other weights.

SHRED YOUR LEAVES

If you're using leaves, it's important to shred them, as whole leaves tend to mat together, blocking moisture and air. Plus, shredded leaves have more loft, which makes them a better insulating material for winter harvesting.

There are a few ways to shred leaves. I use my lawn mower to run over the leaf-covered lawn in autumn. You can use a mulching mower and collect the shredded leaves in a bag or use a standard lawn mower and rake them up afterward. A friend of mine places his leaves in a barrel and uses a string trimmer to quickly shred them into small bits. You can also find leaf mulchers or leaf vacuums that shred at garden supply stores.

Once shredded, bag or gather the leaves and place them in an out-of-the way spot until you're ready to use them. I store several dozen bags of shredded leaves to the side of my garden beds, where they'll be handy when it's time to apply the winter mulch.

VEGGIES
to Mulch for the Winter

- Carrots
- Beets
- Winter radishes
- Parsnips
- Celeriac (celery root)
- Crosnes (Chinese artichokes)
- Jerusalem artichokes (sunchokes)

- Leeks
- Kohlrabi
- Cabbage
- Horseradish
- Parsley root
- Salsify
- Scorzonera
- Rutabaga

You don't want it blowing away in a winter storm. Mark the bed with a 3- or 4-foot-tall bamboo stake. I live in a snowbelt and by late January it can be hard to tell one bed from another. Marking the bed and then labelling the post with a waterproof marker ensures you dig up the crop you want.

Gardeners in Zones 2 to 4 can add a polyethylene-covered mini hoop tunnel over mulched beds for further insulation. When it's time to harvest, lift the side of the tunnel, push back the mulch and pull up your crops. Straw or shredded leaves can also be paired with winter cold frames when root or stem crops like leeks are being grown.

Evergreen branches. Not every winter mulch needs to be a blanket of leaves or straw. Gardeners in Zones 4 to 7 can protect cold-tolerant leafy greens like kale or spinach with a cover of evergreen branches. Use the branches to make a "fort" around mature kale plants or use them to cover young kale and spinach plants in late autumn, before the ground freezes. Remove once the weather is reliably above 40°F (4°C) in early spring. The young plants will begin to send out fresh growth, giving you an extra-early harvest.

A polycarbonate cold frame isn't as insulating as a wood-framed one. Add your own insulation by surrounding a polycarbonate frame with shredded leaves, straw, or evergreen boughs.

CARING FOR OVERWINTERED WINTER CROPS

Winter is a quiet time in the garden when my garden covers and raised beds are usually buried beneath a layer of snow. And while it's definitely a low-maintenance time of year, there are a few things I do to ensure my winter crops stay healthy.

Keep covers clear of snow. Winter structures like cold frames, mini hoop tunnels, greenhouses, and polytunnels capture the warming rays of the sun. In order to do this, they need to be clear of snow. Therefore, as soon as a snow storm has passed, I head out to the garden to brush off any accumulated snow from my garden covers. If it's just a dusting or very lightweight snow, I use a soft broom. Heavy, dense snow or a large accumulation requires a bit more elbow grease. Do not underestimate the weight of snow on a structure. For walk-in structures, carefully brush snow from the roof and sides. To loosen the snow on the top of my polytunnel, I stand inside and tap the roof with a broom. A long-handled car window cleaning brush is also a good choice. Avoid any tools with sharp edges that could tear polyethylene or scratch polycarbonate.

Vent on mild days. Even our frigid northern winters have mild days from time to time.

You can make your own mini hoop tunnels or you can buy them from a garden supply store or online retailer. This one is great for spring and autumn protection, but can also be used into winter.

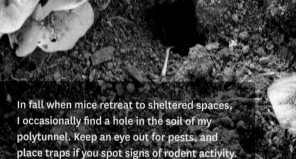

In fall when mice retreat to sheltered spaces, I occasionally find a hole in the soil of my polytunnel. Keep an eye out for pests, and place traps if you spot signs of rodent activity.

Harvest carrots and other root crops before days begin to lengthen.

When the mercury creeps a few degrees above freezing — and particularly if it's a sunny day — I open cold frame tops, lift the ends of my mini hoop tunnels, and crack open a window of my polytunnel. I often double up my garden covers with row cover mini hoop tunnels placed inside my polytunnel. On days where it's very frigid and the inside temperature is below freezing, these covers are left in place. When temperatures rise above freezing, I push back these covers during the day, moving them back over the crops again by late afternoon.

Pest patrol. One of the reasons I love my winter garden is the almost complete lack of pests. I say "almost" because, while there are no slugs or deer nibbling my winter vegetables, over the years I've had to deal with mice from time to time. Good garden hygiene and not leaving any food in the polytunnel (we do occasionally enjoy a meal under the shelter of our tunnel and I've been known to munch a cookie with my cup of tea) are important steps in avoiding rodents. Use cloches or other barriers to keep mice from eating your young seedlings and place traps to reduce the population. And, once the weather begins to warm in late winter, keep an eye out for aphids, slugs, and other common garden pests. (And weeds! Pull any that pop up.)

Harvest. As the days begin to lengthen and winter starts to sprint toward spring, keep an eye on overwintered crops. Any root crops left in your tunnels, cold frames, or beneath mulch should be eaten before they begin to sprout. Same with overwintered greens. They'll soon start to bolt and flavor and eating quality declines.

OVERWINTERING CROPS FOR SPRING

One of the easiest ways to get a jump on the spring harvest is to overwinter vegetables like carrots, parsnips, and spinach. Overwintering hardy vegetables — even in a northern garden like mine — allows you to harvest super sweet roots or tender greens months before the typical growing season even begins.

The planting window depends on the crop. Parsnips are a long-season vegetable and their seed is sown early to mid-spring for a harvest the following spring. Come late autumn, the parsnip bed is deep mulched with straw and the roots overwinter. The following March through May, we dig the still-dormant roots before they begin to sprout. Carrots are overwintered in much the same way,

but because they are quicker to grow, our overwintered carrot seed is planted in early August.

If I have space in my raised beds, I'll also overwinter leafy greens like kale, spinach, mustard, mizuna, tatsoi, claytonia, mâche, and endive. Most of these are direct sown from late summer to early autumn, sprouting quickly in the warm soil at the end of the season. For salad crops I don't overwinter by deep mulching but rather erect a mini hoop tunnel over the seedbed. Once the day length shrinks below 10 hours and winter settles in, the growth of the small plants slows until February. As the light increases in late winter, the overwintered greens begin to grow and the harvesting begins.

YEAR TWO AND BEYOND: ROTATING CROPS

Crop rotation is a basic tenant of organic gardening and just means that you plant a different crop in a given garden bed from year to year. A 3-year crop rotation is a common recommendation and one that I try to stick with in both the open garden and beneath protective covers.

Why 3 years? Because it takes most soilborne diseases and pests at least 3 years to decline to a nonthreat level before susceptible crops can be planted back in that space. It's important to understand that crop rotation isn't about the current growing season, it's a long-term strategy to promote plant and soil health.

Rotating by Edible Part or by Family?

There are many ways to rotate crops, but two of the common ones base rotation on edible parts or on vegetable families.

Edible parts. The method that rotates vegetables by edible parts generally divides them into four groups: leafy greens, fruiting crops, root vegetables, and legumes. The problem with this system is that some related crops (like tomatoes and potatoes, or turnips and cabbage) have different edible parts, but because they're in the same family they're susceptible to many of the same soilborne diseases and pests. When tomatoes follow potatoes, for example, you may be introducing the tomatoes to verticillium wilt or blight that previously affected the potatoes. If you are rotating crops based on edible parts, be mindful of related plants and rotate accordingly.

Family. This method of crop rotation groups vegetable families together and moves them around the garden from year to year. For example, if I grow tomatoes in one of my polytunnel beds I won't plant tomatoes or tomato family crops (peppers, eggplants, potatoes) in that same bed for at least 3 years.

CROP ROTATION

YEAR 1	YEAR 2	YEAR 3	YEAR 4
Tomatoes and Cucumbers	**Root Vegetables**	**Legumes and Greens**	**Cabbage Cousins**
► Tomatoes	► Carrots	► Beans	► Cabbage
► Eggplant	► Beets	► Peas	► Kale
► Peppers	► Radishes	► Lettuce	► Kohlrabi
► Cucumbers	► Turnips	► Spinach	► Broccoli
► Melons	► Onions	► Endive	► Cauliflower
► Squash		► Mâche	► Brussels sprouts
► Potatoes			► Arugula

There are a lot of great reasons to plant vegetable families together. As noted above, related crops often have the same susceptibilities to diseases and insect pests. By rotating them together, you give the pest or disease time to subside before planting that crop family again.

Another reason is soil fertility. Many crop family members also have similar nutrient needs. Planting successive related crops can deplete specific nutrient levels in the soil. An example of this is growing cabbage and then following it with broccoli. These are both in the Brassica family and both require plenty of nitrogen to yield a good crop. If broccoli is planted after cabbage, there may not be adequate nitrogen to form large broccoli heads.

Vegetables that aren't prone to soilborne diseases offer more flexibility in a crop rotation schedule. They can be tucked wherever you have space in gardens and under covers. Examples include lettuce, spinach, beets, endive, and chard, and herbs like basil, parsley, oregano, and thyme.

Under Cover Crop Rotation

With limited space and often fewer types of vegetables grown in garden structures, crop rotation can be a bit tricky, but it's no less important. You must consider crop rotation when coming up with planting plans for polytunnels, greenhouses, and cold frames.

In a walk-in structure, I find it helpful to divide the growing area into separate spaces, even if it's just on paper. Three or four beds make crop rotation a snap, but small structures may not have enough room to accommodate that many beds. In that case pots, fabric planters, or vertical containers can be used as a stage in the rotation plan. For example, in a structure with only two beds, tomatoes can be planted in bed one the first year, bed two the second year, and in containers the third year.

PEAS, PLEASE!

When planning a crop rotation schedule, plant soil-building crops like legumes after heavy feeders like cabbage family plants. Legumes like peas and soybeans are often described as "fixing" nitrogen in the soil; they work with rhizobia bacteria in the soil to convert gaseous nitrogen to usable ammonia.

Have you ever pulled up pea or bean plants at the end of the harvest and noticed tiny nodules on the roots? Those contain the rhizobia bacteria. To keep any excess nitrogen left in the soil at the end of the harvest, dig under your legumes so they can decompose in the garden bed.

SHOULD YOU BUY INOCULANT FOR LEGUMES?

Most garden centers sell packets of what's often labelled as "pea and bean inoculant," which is commercially prepared rhizobia bacteria. The dry, powdery mixture is either mixed with water to coat seeds just before planting or the seeds are premoistened and sprinkled with the inoculant. But do you need an inoculant to grow a bumper crop of legumes? The short answer is "no." I've grown many crops of beans and peas over the years without using a soil inoculant and had decent harvests. However, inoculant does help fix nitrogen to provide a more readily available source to the plants, prompting vigorous growth and a potentially larger harvest. This is something to keep in mind if your garden beds are new or if you're planting legumes in containers under a garden structure.

LEGUME FAMILY

CABBAGE FAMILY

NIGHTSHADE FAMILY

A Guide to
VEGETABLE FAMILIES

CABBAGE FAMILY. Broccoli, Brussels sprouts, cabbage, collards, cauliflower, kale, kohlrabi, mustard greens, radishes, turnips

CARROT FAMILY. Carrots, celery, fennel, parsley, parsnips

GOOSEFOOT FAMILY. Beets, chard, orache, spinach

GOURD FAMILY. Cucumbers, gourds, squash, melons

LEGUME FAMILY. Beans, peas, soybeans

NIGHTSHADE FAMILY. Eggplants, ground cherries, peppers, potatoes, tomatoes, tomatillos

ONION FAMILY. Chives, garlic, leeks, onions

COVER CROPS IN THE UNDER COVER GARDEN

Organic farmers know the value of cover crops, but I rarely see home gardeners include these fast-growing plants in their rotation plans. Cover crops are soil-building superstars, improving the quality, biodiversity, and structure of soil. They also prevent weed growth and soil erosion. In garden-size quantities, the seed is also inexpensive to buy. Why aren't we using more cover crops in our gardens?

For home gardeners, the challenge is time. For example, buckwheat is a popular cover crop that has quick, dense growth and pollinator-friendly blooms. It needs 6 weeks to go from seed to flower, at which point it's cut down or dug under. A further 3 to 4 weeks are needed for the plants to decompose before the bed can be replanted. In total, that means you need to set aside a 9- to 10-week window where you can't grow anything else in that bed.

However, cover crops are super versatile and can easily be worked into a crop rotation plan. Because I have a lot of beds, taking one or two out of production for 2 months isn't that big of a deal, but the payoff is huge.

I generally grow cover crops as "green manures." These are short-term plantings that I dig under to decompose in place and boost soil organic matter. For short-term soil-building opt for quick-growing plants like buckwheat, annual ryegrass, and summer alfalfa. For winter soil building and erosion protection, winter rye is my go-to cover crop.

I even find time to grow a cover crop in my cold frames. I have about 10 weeks between the harvest of the early spring crops and the midsummer seeding of fall and winter carrots. Between these crops, there is adequate time for a cover crop to be sown, grown, dug under, and broken down.

The same applies to my polytunnel. I have two long beds and planting even small sections of the beds with a cover crop from time to time helps keep the soil in this protective environment rich in organic matter and productive.

For me, the biggest cue for when to cut back and dig in my cover crop is when half of the plants are flowering. You don't want to allow the plants to flower and set seed; some cover crops, like buckwheat, can reseed everywhere and become weeds. If the plants are quite tall — some can grow to 3 feet or more — cut them back before digging them under by hand or with a rototiller. They can also be left to decompose on the soil surface or used as a mulch. Winter cover crops often die over the winter and are dug or tilled the following spring.

After using a cover crop, delay the next seeding or transplanting a few weeks to allow the green manure time to start decomposing.

FLUSHING THE SOIL IN POLYTUNNELS

Using chemical-based liquid or granular fertilizers to feed crops can result in a mineral salt buildup in the soil. This is most common with frequent or heavy applications and affects plant health and growth as well as the soil ecosystem. A visual clue to salt buildup is a white crust forming on the soil surface.

To mitigate this problem, avoid overfeeding with synthetic fertilizers, or better yet, go organic! In open garden beds, rain helps flush inorganic products out of the soil. In structures like domes, greenhouses, and polytunnels, where covers are in place for years, there is no rain to clean the soil and periodic deep watering is necessary to remove residues.

BUCKWHEAT

CRIMSON CLOVER

ANNUAL RYE

FIELD PEAS

CHOOSING COVER CROPS

Type of cover crop	When are they grown/planted?	What do they provide to the soil?	How are they incorporated?	Growth rate	Notes
Summer alfalfa	Spring through summer	Fixes nitrogen; deep roots break up soil	Dig or till under when flowering	Fast	Flowers are very attractive to pollinators
Buckwheat	Spring through summer	Loosens topsoil and makes phosphorus available to successive crops	Cut and dig under about a week after plants flower; breaks down quickly	Fast	Flowers are very attractive to pollinators; liable to reseed if not cut down early enough
Field peas	Spring and late summer	Fixes nitrogen	Cut and dig or till under about a week after plants flower	Fast	
Winter ryegrass	Late summer through fall	Prevents winter soil erosion	Dig or till under in spring when plants are a foot tall	Medium	
Annual ryegrass	Spring through fall	Deep roots break up soil	Dig or till under as plants begin to flower	Fast	Seed is quick to germinate
Crimson clover	Spring through fall	Fixes nitrogen	Dig under when plants flower; breaks down in about 10 days	Medium-fast	Flowers are attractive to pollinators

The best time to dig in most types of cover crops is when about half of the plants are flowering. Don't allow them to set seed and self-sow in the garden.

5

Setting Up
SYSTEMS

The systems we use when we garden under cover help us grow healthier plants. Healthy plants grow faster, yield better, and are better equipped to fend off disease and insect infestations. So we capture heat, we ventilate, we provide shade, and we irrigate. It's important to have a basic understanding about how temperature, humidity, and moisture affect plant health so you can provide the best growing environment for your crops.

CONTROLLING TEMPERATURE AND HUMIDITY

Using garden covers is often a dance of heating and cooling, venting and maintaining humidity. We use the sun to heat our cold frames, poly-tunnels, and cloches, and we lift covers, roll-up sides, and open doors and windows to cool and ventilate. The end goal is always the same — to land in the ideal range of temperatures and humidity levels for healthy plant growth.

Temperature Affects Plant Growth

Keeping an eye on the temperature at different times of the day, month, and year helps you provide plants with ideal growing conditions under cover. Air temperature affects the rate of photosynthesis and respiration in plants — that is, how quickly they turn energy from the sun into sugars for growth, and how much water vapor they release from the pores in their leaves. The growth rate of vegetables is determined by the average daily temperature. The growth rate will vary from crop to crop, as each has an ideal temperature range for optimal growth.

I find it helpful to keep a thermometer in my cold frames and polytunnel to monitor temperature fluctuations. I even tuck one in my mini hoop tunnels from time to time to give me an idea of the temperature variation. In fact, I have the display for my polytunnel thermometer on my kitchen counter and I find myself checking and comparing the temperature difference between the outside temperature and the interior temperature of the tunnel constantly throughout the day. Of course,

Ideal Air Temperatures for HEALTHY GROWTH

- ► Beans: 65 to 85°F (18 to 30°C)
- ► Peas: 60 to 75°F (15 to 24°C)
- ► Lettuce: 60 to 70°F (15 to 21°C)
- ► Spinach: 60 to 68°F (15 to 20°C)
- ► Beets: 60 to 70°F (15 to 21°C)
- ► Carrots: 60 to 70°F (15 to 21°C)
- ► Cabbage: 60 to 70°F (15 to 21°C)
- ► Tomatoes: 68 to 80°F (20 to 27°C)
- ► Peppers: 70 to 75°F (21 to 24°C)
- ► Cucumbers: 70 to 75°F (21 to 24°C)
- ► Squash: 70 to 90°F (21 to 32°C)
- ► Basil: 75 to 80°F (24 to 27°C)

the temperature inside the tunnel depends not just on the outside temperature but also on the weather. Sun, clouds, wind, and other meteorological conditions influence polytunnel temperature. Interior temperature can also be influenced by solar collectors like water jugs, which absorb and release heat. On a sunny spring day when the outside temperature is 50°F (10°C), I've recorded the inside temperature of my tunnel at 75°F (24°C). On a sunny August day when it's 79°F (27°C) outside, it can be 95°F (35°C) in the tunnel.

Getting Moisture Levels Right

Humidity — the measure of water vapor in the air — is another important factor in plant growth, and it can be pretty tough to control in a greenhouse, polytunnel, or dome. Too much humidity encourages diseases like botrytis and powdery mildew, and too little slows the growth of plants. The humidity level is dependent on air temperature, as warmer air has a higher capacity for holding moisture than cooler air. When the structure heats up, moisture enters the air as it evaporates from the soil and transpires from plants' leaves. Come evening and night, the temperature goes down and the air releases the excess moisture, which forms a layer on leaf surfaces, glazing materials, and other surfaces in the structure.

When plants are grown under plastic covers like polyethylene and polycarbonate, they don't have access to rain water and soil can dry out quickly, especially on a warm, sunny day. Paying attention to moisture is essential; you need to learn how much water your crops need, how often they need to be irrigated, and when the best time is to water.

INCREASING HEAT RETENTION

Garden structures covered with clear materials like glass, polycarbonate, or polyethylene are focused on light transmission, not heat retention. It's easy to warm a structure on a sunny day, even in cold weather, but the trick is retaining a portion of that heat when the sun goes down and the temperature drops. Using heat sinks to trap and release heat and increasing the insulation in a garden structure helps reduce temperature swings and keeps structures warmer on chilly nights.

Creating Heat Sinks

Heat sinks can be made of any type of thermal mass — material that can absorb, store, and release solar energy. It does double duty by warming in winter and cooling in summer. How? Having a large thermal mass, such as a water tank or rock walkway in a dome or polytunnel, absorbs heat during the day, which cools the structure. That heat is then released at night when temperatures fall.

You can put thermal mass to work in any kind of garden structure, whether large or small.

Walk-in structures like greenhouses, polytunnels, or bioshelters can incorporate thermal mass as a flooring material or in the form of a wall on the north side. For smaller structures like cold frames or mini hoop tunnels, tuck a few jugs of water under the covers.

A layer of bricks in this greenhouse at the Lost Gardens of Heligan, in Cornwall, England, capture heat from the sun.

WATER OR ROCKS?

Certain materials are more effective at storing thermal energy than others are. A material's ability to store energy depends on factors like its density and heat capacity. Water and rocks are common heat sinks in home garden structures, but they're not equally effective. Water holds more heat and absorbs it faster, but it also releases the heat more quickly than rocks do. In addition to rocks and water, materials like concrete, bricks, and even the soil in raised beds act as thermal mass.

Retain Heat while Maximizing Growing Space

When choosing thermal mass for your garden covers, keep in mind that you don't want to add so much that you're using up a good portion of the growing space. And remember that areas directly around the thermal mass will be the warmest spots. Use these spaces for more tender plants.

Water-filled drums on the north side. If you've got space to spare in your walk-in structure, consider adding a few water-filled 55-gallon drums as thermal collectors. Paint them black to maximize heat absorption and place them in full sun where they won't be in your way. Stacking them at the back of a structure, or along the north side saves space and creates a "water wall." If stacking barrels of water, secure them so that they can't fall over and become a hazard. Also, don't fill barrels to the brim; as water absorbs heat, it expands.

Rock or concrete walls. Walipinis and underground greenhouses often use interior rock or concrete walls to capture solar energy.

In-ground heat sinks. Greenhouses, polytunnels, walipinis, domes, and underground greenhouses can be built with an in-ground heat sink in the middle of the structure. Air is warmest at the top of a structure and this air is then forced down pipes by a fan, into an in-ground heat sink that is filled with rocks or other dense materials.

Water reservoirs. A dome, like that of Cam and Andrea Farnell (page 72), captures rainwater and redirects it to a large water reservoir inside their dome. This is used for irrigation but also as a heat sink. In small structures, like cold frames and mini hoop tunnels, smaller scale heat sinks such as 1-gallon water-filled plastic jugs or large stones can be tucked inside to absorb and release heat. Paint plastic jugs black to absorb the most heat.

Stone-lined paths and beds. Using stone flagstones, concrete pavers, or other similar materials for pathways and bed edging inside a structure can help retain heat.

ROCKS VERSUS WATER FOR SOLAR HEAT STORAGE

Material	Cost	Heat transfer characteristics	Size of heat sink	Maintenance
Rocks	Low to medium; rocks are cheap, but constructing a rock wall in a bioshelter can be expensive	Good	Requires a larger heat storage area than water	Minimal; some potential for microbial growth on rocks
Water	Low to medium; water is cheap, but there are additional costs for containers or construction	Excellent; four to five times the heat-storage-to-volume ratio than rocks	Small to medium; practical even in small structures	Medium; potential for leakage or system corrosion

Source: Purdue University

Insulate Structures to Retain Heat

It's important to find ways to increase the insulation in structures like cold frames and polytunnels to trap heat and slow cooling. You can insulate cold frames by lining the inside walls with foam sheets, surrounding them with straw bales or bags of leaves, or banking the outside walls of the frame with soil or mulch. In frigid weather, toss an old blanket on top of a cold frame to trap heat.

To help polytunnels, greenhouses, and domes retain heat in cold weather, check them annually for drafts, making sure windows, doors, and polycarbonate panels fit tight, and seal cracks around doors and windows with weather stripping. Some owners of small greenhouses or domes might choose to add a winter layer of a bubble wrap–like material over the glazing material to retain heat.

TWO LAYERS ARE BETTER THAN ONE

Another option to retain heat is to cover polytunnels with a double layer of polyethylene and fill the space between with a cushion of air. According to Washington State University, an inflated double layer of polyethylene can reduce heat loss by up to 40 percent over glass-covered structures.

You'll often find inflated polytunnels at nurseries but they can also have a place in the home garden. The 6- to 8-inch air cushion is maintained by an inflation fan that runs continuously. Be sure to place the fan so that it draws outside air. Using

the moist air from inside the structure results in increased condensation between the layers.

It's also important to note that inflation fans do make noise and if your structure is in an urban setting or area where you have neighbors nearby, they may not be so thrilled to hear the constant hum of the fan.

Cam and Andrea use insulating covers on the interior north side of their geodesic dome to help prevent heat loss in winter.

Other Ways to MAXIMIZE HEAT in Garden Structures

▶ **Let in the light.** Use a glazing material that allows maximum light transmission. Glass, a single layer of polyethylene, and a single layer of polycarbonate all allow approximately 90 percent of light to enter. Twin-wall polycarbonate or a double layer of polyethylene allows approximately 80 percent of light to enter.

▶ **Get the direction right.** Cold frames, polytunnnels, and greenhouses should be orientated east-west to capture maximum winter light.

▶ **Add row covers.** A layer of row cover fabric acts as a blanket for your veggies. It provides a few degrees of frost protection and is quick and easy to install.

▶ **Cover with a cloche.** A simple cloche made from milk, water, or soda bottles can be slipped over individual plants in a cold frame, greenhouse, or other structure in winter to protect plants.

▶ **Block the wind.** In exposed gardens, wind is a major factor in cooling. Use hedges, trees, fences, garden walls, and other types of windbreaks to reduce wind.

Layering Covers to Retain Heat

It's been many years since I first used garden covers to extend my harvest and grow more food, but I still remember being awed at the effectiveness of a simple row cover. And sometimes two covers are better than one! Layering covers, especially for winter harvesting, traps more heat and offers more protection from cold winter weather.

Doubling up covers inside a polytunnel — adding a cold frame or mini hoop tunnel, for example — is roughly the equivalent of moving your garden a zone to the south. It's allowed me to grow a wider selection of vegetables in the winter, including many that wouldn't survive without that extra protection.

When pairing garden covers, keep track of temperatures using an indoor-outdoor thermometer. With two layers of protection, things can heat up quickly, especially in early spring and autumn, and you may need to remove one layer on sunny days.

COLD FRAMES

Place a cold frame in a walk-in structure like a polytunnel, greenhouse, or dome. You can build a permanent wooden frame, but I find this is where lightweight polycarbonate cold frames come in handy. They're quite portable and easy to move from bed to bed as needed. Keep in mind that a wooden or solid-based frame still needs to face south for maximum light transmission. If the frame box is made from transparent polycarbonate, it doesn't matter.

CLOCHES

Cloches can be placed over individual plants beneath a row cover or polyethylene-covered mini hoop tunnel in spring or fall or in larger structures in spring, fall, or winter. When pairing cloches with other types of covers, I generally use clear cloches like glass jars, soda bottles, or large water bottles. Opaque milk jugs will do in a pinch, but they need to be removed during the

Sometimes two covers can be better than one! I layer my covers, using cloches or mini hoop tunnels inside my polytunnel. This is especially effective in early spring, late autumn, and winter when temperature fluctuations can be extreme.

day to allow light to reach the plants. You can also buy water cloches which are flexible plastic cone-shaped cloches that have ribs you fill with water. They offer more insulation than glass or plastic cloches, as the water absorbs heat during the day and releases it slowly at night.

MINI HOOP TUNNELS

Quick to construct and so versatile, I've seen mini hoop tunnels placed over winter cold frames, but they're more commonly paired with greenhouses, polytunnels, and domes. Remember to vent often in spring and autumn, and even in winter when the temperature is mild inside your structure.

ROW COVERS

Float lightweight fabrics on hoops or lay them on top of plants beneath a mini hoop tunnel or in larger garden structures. Remove when the sun is shining and the inside temperature is above 40°F (4°C) to allow air to circulate around plants. I only lay covers directly on top of crops as a temporary layer of protection. Leaving the fabric on the foliage for more than a day or two can result in stunted growth or disease issues. Plus, if a row cover is in contact with plants in freezing temperatures, it can damage the foliage. I use row covers as an extra layer of protection in my polytunnel for much of the winter, floating the fabric above the bed on 9-gauge-wire hoops.

We added a string of LED
lights to our polytunnel to
extend our enjoyment of this
space after sunset. Of course,
lights can also be used to
provide additional heat in
greenhouses, polytunnels,
and even mini hoop tunnels.

PROVIDING ADDITIONAL HEAT

If you've sited your garden structures near an outdoor outlet, you'll have a few other options to provide heat, including heating cables and string lights. These can be put to work in spring and fall to protect crops from frost, or you can use them during the winter to provide enough heat for cold-season crops to make it through periods of extreme cold.

Heating Cables

Heating cables are underground electric cables that are woven in the soil of a raised bed or cold frame to warm the soil. They're often used in seed starting to improve germination but can also be sunk into the soil of a cold frame to create ideal conditions for spring, autumn, and winter crops. I don't use heating cables in my frames, but I am tempted to run a section in one of my polytunnel raised beds to create a propagation bed where I could get a head start on starting seeds for transplants. A layer of row cover paired with the heating cable would create a cozy spot for establishing seedings.

There are various sizes of cables available, but a general rule is to buy 4 feet of cable for each square foot of soil. So, a 4-by-3-foot cold frame with 12 square feet of growing space needs a 48-foot heating cable.

Many cables also come with a built-in thermostat that turns on the heat when the soil cools to a certain temperature. On warm or mild days, the cable stays off. On cold days, keep the cold frame top shut to help retain heat. In a larger structure, like a polytunnel, cover the heated portion of the bed with a mini hoop tunnel or row cover. Beds heated with a cable dry out faster, so keep an eye on soil moisture and water when necessary.

INSTALLING A HEATING CABLE

To be most effective, a heating cable should be installed before you fill a cold frame or bed with soil. Be sure to read the manufacturer's instructions before installing one. They generally recommend cables be placed on several inches of sand and then covered with 2 more inches of sand. When laying it out, the cable loops should be about 3 inches apart, never touching or crossing over (this can cause the cable to overheat and stop working). Add a section of hardware cloth (metal mesh) to the top layer of sand so that you won't damage the cable with future digging or cultivating. Finally, add 3 to 4 inches of soil. If necessary, the cables can also be buried directly in soil. Ensure that no part of the cable is above the soil.

Let There Be Lights

Using string lights or holiday lights to supplement heat in a cold frame, mini hoop tunnel, or other small structure does offer a measure of success. An experiment conducted by Colorado State University found that a string with 25 C7 (medium-size) incandescent lights provided between 6 to 18°F (3 to 10°C) of frost protection in a cold frame. The cold frame they tested measured 4 by 5 feet and the lights were turned on at dusk and off at dawn.

Keep lights away from foliage and the polycarbonate top of a frame or polyethylene covering a mini hoop tunnel. Also, be sure to use an exterior-grade extension cord to power the lights. Check the light cord periodically for damage or wear and tear, replacing it if you spot any issues.

Whether your garden cover is large or small, you must consider ventilation. Good air flow reduces temperature buildup, but it also reduces condensation, helps foliage dry off, and decreases the risk of many common plant diseases.

VENTILATION

One of the main jobs of garden covers is to capture heat and create a microclimate around plants. But you'll find there are times in spring, summer, and autumn (and sometimes even during a midwinter thaw) when the temperature soars and you need to ventilate your crops. All types of covers, including row covers, cloches, greenhouses, and tunnels, need to be vented for good air flow and to prevent major temperature fluctuations, which are stressful to plants.

Maintaining too warm a temperature in spring and fall encourages soft green growth, which is more susceptible to damage when temperatures drop. Giving covered crops tough love and growing them in cooler temperatures results in more resilient plants. Exposing plants to wind that ruffles their foliage encourages sturdier growth. Increasing air flow also lowers humidity levels (thus reducing the risk of disease) and provides fresh air and CO_2, which is necessary for photosynthesis and healthy crop production.

Ways to Vent

Smaller covers. In small season extenders like mini hoop tunnels or cold frames, open the ends or lift the tops to vent. I use binder clips to hold up the ends of the polyethylene or fabric on my mini tunnels, but you can also just fold them up if it's not windy. I've used many different

Reduce humidity by venting well and encouraging good air flow, not watering late in the day, and not leaving water to puddle in beds or pathways.

props to hold my cold frames open — sticks, logs, rocks, bricks, buckets, and so on — but a strong stick is my preferred prop. You can buy or build a notched cold frame prop to hold the tops open to varying degrees. Cloches will also need to be vented on mild days. If the temperature allows, they can be removed completely and set aside until evening, or you can tuck a small stone or wedge of wood under them to allow air to flow. Some cloches are self venting. With 1-gallon milk jugs, for example, just remove the tops to let hot air out. Purchased plastic cloches often have vents on top that can be twisted opened or closed as needed.

Larger structures. Greenhouses and domes can be constructed with pop-out windows for ventilation and cooling. It's more difficult to install roof vents in polyethylene-covered polytunnels, so gable-end fans and roll-up sides are used to ventilate the structure. When planning a polytunnel or greenhouse, keep in mind that size affects the rate of cooling. It's harder to adequately vent structures wider than 20 feet and longer than 48 feet without roll-up sides or a ventilation system. Opening doors at each end won't be enough to move air through effectively.

A circulating fan can keep air moving in a greenhouse or polytunnel. There are electric fans

Manual or Automatic?

Ventilation can be manual or automatic. Automatic vent openers can be purchased online or at garden and greenhouse supply stores. They can be installed to open cold frame tops or greenhouse windows. Most contain a cylinder of temperature-sensitive fluid that expands and contracts as the temperature rises or cools, opening and closing the structure.

Although they're pricey, automatic venting systems are convenient, especially if you aren't able to vent your garden covers by hand when it's needed (like at midday, when you're at the office). Automatic options can save money in the long run. In today's high-tech world, there are many venting and monitoring systems that can be controlled from a smartphone or laptop.

When in Doubt, Vent

I'm a big believer in erring on the side of venting. If the outside temperature is forecast to be over 40°F (4°C), I crack open the cold frame tops, untie the ends of mini hoop tunnels, and open the windows and door on my polytunnel. If the sun is out and the temperature is above 46°F (8°C), I open frames completely and roll up the sides at least partway on my polytunnel.

An automatic venting arm ensures that Cam and Andrea's dome stays well ventilated.

as well as solar powered ones that run when the sun is shining. The size of the fan will depend on the size of the structure. Talk to your structure manufacturer to find the right size fan, or try one of the fan calculators available online. Before investing in a fan, though, I'd suggest trying the other cooling techniques first.

Other Ways to
COOL A STRUCTURE

THINK SHADE. Shade cloth is a quick, easy, and inexpensive way to cool a garden structure. Shade paint (a mixture of water and white latex paint) is a traditional way to shade and cool a glass greenhouse, but it's labor intensive to apply and tedious to remove.

THERMAL MASS. To warm, thermal mass is placed in the sun, but to cool, thermal mass should be located in a shady spot such as alongside beds planted with tall or vining crops or behind staging or a potting table. It absorbs heat during the day, cooling the interior of a greenhouse, dome, or polytunnel.

WATER THE PATHS. Adding water to the interior of a structure raises the humidity, which makes it easier for plants to deal with high temperatures. Spray cold water on pathways several times a day if you happen to be home.

IRRIGATION

Growing under cover means that rain often cannot reach your crops. With permeable covers like shade cloth and row cover, irrigation isn't necessary unless there has been no rain. And if the temperature is mild, small garden structures like cold frames and mini hoop tunnels can be opened to allow rain to reach crops. But in larger, more permanent structures, you'll have to devise a plan to irrigate your crops effectively.

Watering Well Under Cover

You'll need to supply the water when gardening in greenhouses, polytunnels, or domes. No gardener wants to be stuck watering every day — or several times a day in hot weather — so keep irrigation in mind when deciding on a location for your structure, choosing a spot with a water source close at hand if at all possible.

How often you'll need to water depends on a number of factors, including the time of year and the soil type. On a hot, sunny summer day, the soil will obviously dry out quicker than it would on a cloudy spring day. At the opposite end of the season, when the day length shortens and temperatures drop in autumn, watering slows again. Common sense does come into play and you'll soon develop a feel for how often you need to water. Placing a thermometer inside your garden covers is a great way to keep an eye on the temperature and track how quickly water may be evaporating.

Understanding your soil type is crucial to figuring out how often to water. Water moves quickly through sandy soil, whereas loamy soil that's rich in organic matter will hold water better and need to be irrigated less frequently. Water will pool on the surface of clay soil if it's overirrigated.

Watering at the right time, in the right way, and providing the right amount of water is especially important when plants are grown under cover. A rain barrel helps deliver water to where the irrigation doesn't reach.

WATER SMART!

Water makes up 80 to 90 percent of a plant's weight, so having adequate, consistent water is critical to healthy plant growth. Uneven or inconsistent watering leads to water stress, which makes crops more prone to bolting or physiological conditions like blossom end rot. Here are a few tips for watering well under cover.

Water at the right time. If possible, water early in the day so that foliage has a chance to dry before nightfall. Wet foliage promotes disease.

Water before plants wilt. Don't wait for your crops to tell you they're thirsty. Once they wilt they're water stressed, and that affects plant health, growth, and productivity.

Don't overwater. Overwatering is as bad as underwatering. Too much water can result in wilted leaves (yes, leaves wilt when they have too little *and* too much water), yellowing leaves, or root rot (when the roots turn soft and mushy). Not sure if you're watering too much? Before you grab the hose, stick your finger in the soil. If it's still moist, you don't need to water. If it's dry 2 to 3 inches down, it's time to irrigate.

Water deeply. Avoid frequent, light irrigation. This encourages a proliferation of shallow roots.

Pay attention to differing water needs. Certain crops are thirstier than others. Use soaker hoses to deep irrigate thirsty vegetables or, if hand-watering, be mindful to water them more frequently.

Mulch to conserve water. Soil that is shaded with mulch needs less watering. Mulch crops with a natural material like organic straw, shredded leaves, spray-free grass clippings, or leaf mold compost.

Water with the Weather

Anyone who has gardened in a greenhouse for any amount of time will tell you that every irrigation session isn't necessarily the same. On cloudy or rainy days, plants use less water and you'll need to adjust your watering accordingly. I often don't water on rainy days. If I must, I'm even more careful than usual about wetting the foliage as it likely won't dry quickly when the weather is cloudy or damp.

Harvesting Rainwater

Rainwater has some advantages over tap water, particularly if you're on a municipal water source. Municipal water often contains chemicals, salts, and minerals like chlorine and fluoride, which can damage plants. Tap water is also much colder than collected water and can potentially shock tender seedlings and cold-sensitive plants. If you're using tap water, fill containers and allow them to warm up before watering plants.

If you'd like to harvest rainwater, why not use your polytunnel or greenhouse to help collect it? Many structure manufacturers sell gutter brackets and gutters, or you can create your own system. Rainwater can also be collected from the roof of a house or other nearby structure. For your own convenience, water barrels should be located as close as possible to your garden covers.

IRRIGATION TECHNIQUES AND SYSTEMS

Which irrigation technique you choose depends on how large your covered garden space is and how much time you're able (or want) to spend watering by hand. It's easy to water a cold frame or two with just a watering can, but a large polytunnel may need a more automated system.

Hand Watering for Small Structures

Hand watering is the most common method of irrigation in home gardens and small structures — whether that means using a watering can or watering by hand with a hose-end sprayer. Watering by hand gives you an opportunity to visually evaluate your crops, checking growth and health. The downside is that hand watering is more time consuming than using an automated system, especially in summer when you may need to water more than once a day.

I've hand-watered my gardens, mini hoop tunnels, cold frames, and polytunnel for years and find that a watering wand makes quick work of this task. A long-handled watering wand helps me water the base of the plants without wetting the foliage. Plus, hand watering allows me to be precise. The extrathirsty crops get plenty of water and the drought-tolerant ones get less.

Dragging a hose around a walk-in structure can be a hassle. It gets dirty, kinked, and can accidentally squash or damage plants, or — even worse — knock unripe tomatoes or peppers from your plants! If you can, run your hose overhead and support it on crop bars or trusses. Or use a hose trolley in your greenhouse or polytunnel to keep it tidy and under control.

They may seem clunky, but don't forget about old-fashioned watering cans, which are perfect for spot irrigation. I use my watering cans all the time, leaving them filled up just inside the door of my polytunnel or beside the cold frame.

To reduce the need to water by hand, you can set up an irrigation system. A soaker hose and drip tape deliver water directly to your crops. Avoid overhead watering systems, which spray water everywhere and can increase the risk of plant diseases.

Using Soaker Hoses

Soaker hoses are a convenient and efficient way to water crops in mini hoop tunnels, polytunnels, greenhouses, or domes. They're usually made of recycled rubber, with tiny pores that weep out water along their entire length. Typically, hoses are laid beside the stems of the plants — either on top of the soil or just beneath the surface — allowing for efficient irrigation without waste. The slow, deep release of water in direct contact with the root zone means less water is lost to runoff or evaporation. Having the water at the base of the plants means foliage stays dry, reducing the risk of fungal diseases. Another bonus is that the soil stays drier between plants, discouraging weed growth.

INSTALLING SOAKER HOSES

Soaker hoses come in various lengths but are most often sold in lengths of 25 to 50 feet. Going with a longer hose can result in uneven watering, which is a common issue with soaker hoses. If you're having trouble maintaining water pressure all the way down the hose, consider using several shorter hoses instead of one long one. Use a Y splitter at the water source to attach the hoses.

It's also important to use soaker hoses on level ground. If the water has to flow uphill, pressure loss is inevitable.

Use a regular garden hose to run from the faucet to the soaker hose. I'd suggest installing a pressure regulator on the faucet to slow down the flow of water or just turn the faucet on part way. Too much water pressure can result in tears, leaks, or bursts. If you garden in an area with hard water, it's also a good idea to add a calcium filter to eliminate mineral buildup in the hose.

When laying soaker hose, try to cover as many plants as possible. This usually means weaving and circling the hose through the bed and around your vegetables. Keep hoses in place with wire garden pins.

GAUGING WATER PENETRATION

After installation, pay attention to how long it takes to water the soil deeply. Dig into the soil to see how far down the water has reached. This will help you figure out how often you need to water and how long you need to leave the hose on. As the weather heats up in summer, expect to water more often. To automate the system, add a timer.

A soaker hose can be run around the base of plants to deliver water right to the root zone. It can be laid on top of the soil, buried slightly, or topped with a layer of straw mulch.

Watering in the
SHOULDER SEASONS

Watering from mid-spring through late autumn is a piece of cake. I've got my hose handy and a few watering cans for those far-off spots that my hose doesn't reach. Yet in the shoulder seasons of late autumn and late winter to early spring, watering can be a challenge. Why? Cold weather can freeze the water in your hoses overnight, which means you need to unhook and drain them every time you use them. I also get asked a lot about watering in the winter. It's very rare that I need to water then; the exception would be if we have a February thaw and I'm sowing seeds in my cold frames or polytunnel.

The good news about watering in early spring and late autumn is that you don't have to water nearly as often as you do from mid-spring to mid-autumn. Plants are growing slower and using less water. Plus, the outdoor temperatures are cooler, which means less transpiration, so soil dries out less frequently. If you are going to water, do so early in the day so the leaves have plenty of time to dry off before night.

So, what are the options for watering during the shoulder seasons? In walk-in structures, you can use a large container like a 55-gallon barrel as both a heat sink and a source of irrigation water.

Dip a watering can or insert a small submersible pump with an attached hose into the barrel to make watering a snap! You can, of course, use a hose attached to an outdoor water source, but if the night temperature is forecast to dip below freezing, you'll need to drain the hose.

Another thing to keep in mind is that plants don't love a cold shower. Icy cold water won't do your plants (especially seedlings) any favors and can cause shock and stress. I irrigate young plants with a watering can during the shoulder seasons, using room temperature water. Unfortunately, if you're using a hose, cold water is all you've got. You can use the hose to fill a barrel in your greenhouse, let it warm in the sun, and then water your plants with this water. One gardener told me he keeps an aquarium heater in his greenhouse water barrel to heat the water during the shoulder seasons.

If you live in a snowbelt like me, you can also shovel snow directly onto empty tunnel, dome, or greenhouse garden beds in late winter. It will melt and soak the soil. This is also when I like to top-dress the beds with compost or aged manure in preparation for planting.

Overhead Sprinklers

An overhead sprinkler system waters everything in its path: foliage, flowers, fruits, and soil all get soaked. Because of that, it's my least favorite way to irrigate; splashing water around a structure is an easy way to spread disease. Overhead sprinkling also has a much higher rate of evaporation than watering with a soaker hose or carefully hand-watering with a wand. In addition, plants all get the same amount of water, which isn't ideal in mixed plantings. Newly planted seeds and seedlings require less water than more mature crops or those starting to flower or fruit, and overhead watering doesn't account for this.

Overhead sprinkler systems, however, are common in commercial greenhouses, and some polytunnel kits include overhead irrigation systems as an option. If you decide to go this route, turn the system on early in the day to give foliage plenty of time to dry off before nighttime and to reduce the amount of evaporation.

Automating Your Irrigation Setup

If monitoring your irrigation setup gets to be too much of a chore, or if you plan on being away from home for an extended period of time, you might consider installing an automated system. Most systems involve a battery-operated timer attached to a main water supply. You can also hook an automated system to large barrels or reservoirs inside or outside your structures. If using barrels, you'll need to rely on gravity to create adequate water pressure. Raise the barrels several feet in the air on a deck or platform made from wood or cinder blocks and then direct water through hoses and connectors to garden beds. Or you can use a submersible pump (preferably solar powered) to pump water from water barrels or reservoirs to your crops.

An irrigation timer ensures that your crops are watered, even when you're away.

Watering and Fertilizers

Be mindful that if you're using a topdressing of granular fertilizer to feed your plants, the water from the soaker hose, watering can, watering wand, or sprinklers can displace the fertilizer, washing it away from the plants. To avoid this issue, scratch granular fertilizer into the soil around the base of your plants or use liquid organic fertilizers during the growing season.

VACATION OR STAYCATION?

What do you do when you want a few days or few weeks away from your garden covers but need to ensure plants are properly irrigated? You can call a garden-loving friend to water or, if you're technology savvy, you can set up a system that can be monitored from a smartphone. This allows you to make irrigation (and venting) adjustments as needed — no matter where you are in the world. If an automated system isn't in your budget or doesn't fit into your garden, here are a few tips to help keep your plants watered for as long as possible:

► Water just before you leave, giving everything under your garden covers a deep soaking.

► Make your own DIY drip irrigation with half-gallon plastic bottles. Clean the bottles and then poke holes in the caps. Fill with water and place upside down near crops in your structures. They will slowly drip water into the beds, giving you a few days' grace.

► Mulch the soil surface with an inch of grass clippings or several inches of straw or leaf mold compost.

► Throw a few hoops over your garden beds or cold frames and top with shade cloth to minimize water evaporation from the soil.

6

Preventing DISEASES, PESTS, and POOR POLLINATION

Although I strive to provide optimal growing conditions in my garden structures, things don't always go according to plan. Sometimes I find slugs nibbling on my spring lettuce or the midsummer cucumbers are hit by powdery mildew. It's also important to keep in mind that it can be just as important to invite pollinating insects into your covered garden as keeping destructive ones out.

PREVENTING DAMAGE FROM INSECTS AND DISEASES

Unfortunately, covers like domes, cold frames, and polytunnels that shelter crops from cold and bad weather don't always protect them from common garden diseases and pests. Diseases can come into your structures via infected seeds, plants, soil, tools, insects, and wind. Insect pests often enter when garden covers are being vented, through small cracks around doors and windows, or when new seedlings and plants are being introduced to the structure. However, there are ways to prevent or minimize the damage they cause.

Tidy up. Keep garden structures free of crop debris and clip dead leaves immediately. In fact, as the summer season progresses, I regularly clip blemished leaves from the tomatoes, peppers, cucumbers, and melons in my polytunnel. Generally, these are the bottom leaves but they can encourage the spread of diseases like botrytis or provide shelter for pests like slugs.

Keep it clean. If you have disease or pest problems in your open garden, be mindful about transferring them into your covered garden. Consider keeping a slip-on pair of garden shoes or boots just inside walk-in structures, to wear only inside polytunnels, greenhouses, or domes. It's also a good idea to have a set of tools that are only used in these structures, cleaning them often to prevent the spread of common diseases.

Ventilate. Venting small and large garden structures is critical to healthy plant growth. It decreases inside temperatures, reduces humidity, and improves air flow. Good air circulation helps foliage dry quicker, which is important in disease prevention.

Be vigilant. When you're seeding, transplanting, watering, weeding, or just generally puttering around your garden structures, take a peek to see if you spot anything out of the ordinary. Look at new shoots, leaf surfaces (above and below), stems, flowers, and fruits. Catching an infestation or outbreak early makes a huge difference.

Water smart. Watering smart means watering at the right time of day, not over- or underwatering, and watering the soil, not the plants. See page 123 for more information.

My polytunnel has roll-up sides which permits pollinators and beneficial insects to come and go. I also include flowers in my tunnel, to attract good bugs and keep pest populations low.

GARDEN DISEASES TO KNOW

Below you'll find some of the common diseases that can afflict under cover vegetables. While sometimes the occurrence of a disease is out of your control (I once got late blight from spores blown up to Nova Scotia in a tropical storm!), generally your cultural practices, dedication to crop rotation, good garden sanitation, and variety choices can greatly reduce the appearance of vegetable garden diseases. Of course, growing under cover does change the gardening landscape in that you're growing plants in an unnatural environment that can be hot and humid, which are favorable conditions for many diseases.

One of the reasons I like to hand-water my garden beds and structures is that it gives me a chance to keep a close eye on my plants and spot developing problems. Vigilance and early identification are key when dealing with a problem. Knowing what type of disease you've got can help you tackle it effectively and prevent future occurrences.

Powdery Mildew

WHAT IT IS: Powdery mildew is the name given to a number of fungal diseases which generally look and act in similar ways. The identifying characteristic is the growth of a white-gray powdery substance on the foliage, buds, flowers, and fruits of plants.

POWDERY MILDEW

WHAT IT DOES: Powdery mildew rarely kills plants, but it does look unsightly and reduces photosynthesis. Affected leaves turn yellow and eventually fall, weakening the plant. Depending on the type of vegetable and the point in its lifecycle, powdery mildew may affect bloom and fruiting, reducing harvest.

HOW TO GET RID OF IT: Powdery mildew is more likely in stretches of damp or humid weather, but it can be encouraged by insufficient air circulation. Be sure to plant seedlings at a proper distance to each other (don't overcrowd!) and ventilate structures to allow air to flow. Keep an eye on susceptible plants like squash and cucumber, removing any leaves that show signs of powdery mildew. Avoid watering from above as splashing water can further spread spores. Spray organic fungicides to prevent or control powdery mildew. Many sprays are made from sulfur, but you can also make your own with ingredients like baking soda or milk. In fact, I've been experimenting with a milk-based spray to prevent powdery mildew over the past few seasons and have been surprised at its effectiveness. I mix one part milk to two parts water in a clean hand sprayer and dose the top and bottom surfaces of my cucumber, squash, and pumpkin leaves weekly.

Botrytis (Gray Mold)

WHAT IT IS: Botrytis is a gray mold that can affect a wide variety of vegetable plants. Cool, humid temperatures (a typical Nova Scotian spring!) and a lack of good air flow in structures are all conducive to the development of botrytis.

WHAT IT DOES: This common disease can infect leaves, stems, flowers, and fruits, and develops into a gray-brown rot that quickly covers the affected parts in a fluffy mold. Mushy and gross! I find botrytis is most common on my container-grown strawberries in late spring when humidity is high and temperatures are cool.

BOTRYTIS

BLOSSOM END ROT

HOW TO GET RID OF IT: Keep your garden structures free of crop debris and be sure to pick ripe fruit promptly. Open doors, covers, tops, windows, and roll-up sides to improve air flow, especially when conditions are cool and humid. Clean pruners often, as they can be a vector for spreading the disease. This is especially important when pruning back crops like tomatoes. Avoid overfertilizing. You may think you're doing your plants a favor by feeding often, but your goal should be to encourage healthy, balanced growth. Overfed plants with plenty of lush, soft growth are more susceptible to diseases like botrytis, as well as to infestations of insects like aphids.

Blossom End Rot

WHAT IT IS: It's heartbreaking to spend months growing tomatoes or peppers only to discover large, leathery patches on the underside of the fruits just when they're approaching maturity. This isn't a plant disease, but a physiological condition caused by a calcium imbalance. Calcium uptake can be affected by several factors, including inconsistent soil moisture or a low soil pH.

WHAT IT DOES: Once it appears, that sunken, discolored patch at the blossom (bottom) end of the fruits continues to spread, covering up to half of the fruit. Affected fruits are unsightly, but still

edible if they've matured enough for harvesting. Cut out the rot.

HOW TO GET RID OF IT: It's much easier to prevent blossom end rot than try to cure it. Blossom end rot is most common in container-grown tomatoes, where inconsistent watering prevents calcium uptake. If growing in pots or fabric planters inside garden structures, water evenly and water often. Never let tomato plants wilt. Soaker hoses or drip irrigation can be installed to regulate irrigation in the garden and with potted tomatoes.

If blossom end rot appears on the early fruits of indeterminate tomatoes, there's a chance you can do damage control and still end up with a late season harvest. Make sure you water consistently, and mulch plants with compost or straw to hold moisture. If this is an annual problem in your garden and covers, do a soil test to see if your soil is calcium deficient. Lime or ground up eggshells can be added to garden soil to boost calcium.

Late Blight

WHAT IT IS: Late blight is a cruel disease, and the one responsible for the Irish potato famine in the mid-nineteenth century. One day, your tomato and potato plants look fantastic and the next, they're a collapsed mass of watery brown leaves.

WHAT IT DOES: This pathogen is wind-borne and can travel miles to infect new plants. The first signs of late blight are brownish lesions on the stems and leaves. The disease moves fast under ideal conditions and a white fuzzy growth develops on the lesions. Soon the plants collapse and die.

HOW TO GET RID OF IT: Late blight thrives in conditions with high moisture and moderate temperatures (60 to 80°F/15 to 26°C). As soon as it's identified, remove all parts of affected plants to curb the spread. This includes any potato tubers developing in the ground. Do not add any part of the diseased plant to your compost pile. Throw them in the garbage. I always grow a few "just-in-case" late blight–resistant tomato varieties like 'Jasper', 'Defiant', or 'Mountain Merit'.

Early Tomato Blight

WHAT IT IS: A fungal disease that is among the most common afflictions of tomatoes grown in home gardens. In my region, early blight is an annual problem for most gardeners, but I've found good growing practices to be key in reducing or eliminating the occurrence of early blight.

WHAT IT DOES: Early blight shows up as roundish yellow-brown spots on the surface of tomato leaves. They can grow up to ½ inch across and appear to have a "target" or concentric circles. The fungus appears on the lower leaves first, working its way up the plant. It also affects stems and fruits.

HOW TO GET RID OF IT: Prevention is key to minimizing the impact of early blight. First, I make sure to rotate my tomato plants on a 3- to 4-year rotation cycle. When I plant the seedlings in the open garden or under garden covers in late spring, I immediately cover the soil with a mulch of straw or shredded leaves. As the plants grow, I clip the lower leaves to limit the ability of the fungal spores to reach the plant. I'm also very careful about watering and try never to wet the

LATE BLIGHT

foliage. There are also varieties like 'Mountain Magic', 'Defiant', and 'Jasper' that show excellent resistance to early blight.

Leaf Mold

WHAT IT IS: Leaf mold is a disease that thrives in high humidity (over 85 percent relative humidity), making it more common for tomatoes grown under cover.

WHAT IT DOES: While the plants may become weakened and the eventual harvest diminished, plants that are kept otherwise healthy can generally be expected to produce a decent harvest. The first indication of leaf mold is a pale yellow discoloring on the foliage of the bottom leaves. This sometimes is mistaken for a nutrient deficiency, but soon moldy-looking brown patches appear on the top and bottom of the leaves.

HOW TO GET RID OF IT: The first line of defense is to grow resistant varieties like 'Geronimo' and 'Bellini'. It's also important to give plants as much air circulation as possible by spacing them properly and pruning regularly. Vent structures well and never allow affected debris to remain in the structure. The fungal spores can persist in the soil for a year or more, so rotating tomatoes with unrelated crops in tunnels, domes, and greenhouses is vital.

SLUGS

APHIDS

WHITEFLIES

UNDER COVER PESTS

The best way to prevent pests under garden covers is to promote healthy plant growth and maintain a balanced ecosystem. Healthy plants have plenty of light, water, and nutrients and aren't too cold or too hot. A plant that is in good health is less susceptible to pests, and is better able to withstand an attack.

As for a balanced ecosystem, I plant my polytunnel and other covers much the same way I plant my garden: with a diverse assortment of plants — vegetables, herbs, and flowers. I keep crop debris to a minimum and vent often to prevent heat buildup, keep air flowing, and allow beneficial insects access to my crops.

Also, buy smart and check first! If you're purchasing seedlings from a garden center, garden club sale, or a big-box store, give the plants a good once-over to check for hitchhiking pests. Once they're established inside your garden covers, it can be tough to eliminate pests.

Slugs

WHO THEY ARE: Slugs are among the most common garden pests and can quickly devastate young seedlings as well as established crops. These land-based mollusks leave a slimy trail wherever they slither.

WHAT THEY DO: In my garden, slugs are a major issue in early spring and autumn when temperatures are cool and there is typically more moisture. They lurk beneath foliage, rocks, or other debris during the day, emerging at night or on cool, cloudy, or wet days to feed. Once temperatures rise inside my polytunnels, mini hoops, and cold frames in mid-spring, the slugs tend to disappear.

HOW TO GET RID OF THEM: Handpick them and toss them into a pail of soapy water. I should become a professional slug picker — I can gather a container of slugs from my garden beds and covers in record time. I like to go out in the early morning, when the slug population, which thrives in cool, damp conditions, is at peak, using gloves to pluck slugs from the soil and plants. Occasionally I come across a clump of slug eggs, which resembles a pile of tiny crystal balls, and I dispose of those, too. I've also had good luck surrounding slug-prone seedlings like pole beans and cucumbers with a layer of diatomaceous earth to protect from slug damage.

Aphids

WHO THEY ARE: One of the more common garden pests, aphids are soft-bodied insects that suck the sap from plant tissue.

WHAT THEY DO: Aphids often cluster on tender shoots and flower buds, causing foliage to distort as it grows. They reproduce very quickly and an infestation can happen almost overnight. One of the first signs of a serious aphid infestation is the appearance of sooty mold, a black mold that grows on sugary honeydew, an aphid secretion. Sooty mold looks messy and gross on plant foliage and the honeydew attracts ants.

HOW TO GET RID OF THEM: In the garden, I remove aphids by knocking them from plants with a hard jet of water from my hose. Once they fall to the ground, they either starve and die or get gobbled up by ground beetles and other predators. If you've discovered aphids under garden covers like cold frames, mini hoop tunnels, polytunnels, or greenhouses, this technique can still work. In serious infestations, an organic product like insecticidal soap makes quick work of aphids. If there aren't too many aphids, remove them by wrapping your hand in sticky tape, sticky side out, and gently picking up the aphids with the tape.

Whiteflies

WHO THEY ARE: Although they look like tiny moths, whiteflies are actually flies and a frequent pest in greenhouses and polytunnels. They're commonly found clustered on the undersides of pepper, tomato, or cucumber foliage, flying into the air when disturbed.

WHAT THEY DO: Whiteflies have a lot in common with aphids; both feed on plant sap and produce a sticky honeydew that encourages the growth of sooty mold. Both aphids and whiteflies can also spread plant diseases as they move from plant to plant sucking sap.

HOW TO GET RID OF THEM: Like aphids, whiteflies can be knocked off the bottom of the leaves with a hard jet of water. Once they fall, whitefly nymphs can't climb back up the plants. You can also spray the colonies with an insecticidal soap. Several applications is usually enough to reduce or eliminate the population.

Cabbage Worms

WHO THEY ARE: If you're a lover of cabbage family crops like kale, broccoli, cabbage, and cauliflower, chances are good that you're familiar with cabbage worms. This widespread pest is sometimes hard to spot because its larva form is a small green caterpillar that blends in with the foliage. I often spot them camouflaged on the underside of kale or broccoli leaves, happily munching away. The good news is that garden covers were made for preventing pests like cabbage worms from decimating your cabbage vegetables!

CABBAGE WORM FRASS

NASTURTIUM

CHIVES

MARIGOLDS

ZINNIAS

ALYSSUM

LACEWING

Bring in the
BENEFICIALS

Don't wait until you've got a problem to think about good bugs. Keep them in mind way back in winter when you start to plan your garden. Consider what types of pests you've encountered before — and are, therefore, likely to see again — and how you might be able to prevent them. In general, the best way to discourage pests is to grow healthy plants and encourage beneficial insects like ladybugs and lacewings. Here are a few ways to welcome beneficial insects to your garden.

Stay away from sprays. That means you need to avoid spraying pesticides, even organic ones, as they affect both bad and good bugs. This can be a challenge for gardeners, especially when faced with an aphid infestation, but waiting a few days and giving those ladybugs time to find the pests generally pays off.

Add flower power. Beneficial insects don't just eat bugs. Many also need the sugars found in plant nectars. Growing a diverse array of flowering plants with different flower sizes, shapes, and colors will attract the widest variety of beneficials. Good flowers to plant include yarrow, sunflowers, dill, zinnias, nasturtiums, and cilantro.

Should You Buy Beneficial Insects?

There is some debate about whether introducing nonnative beneficial insects to our gardens is a good idea. For example, ladybugs available for purchase are often wild-caught in the mountains of California during their hibernation period. In most parts of North America, these are nonnative species, which could carry diseases or parasites that would affect native species. If you decide to buy beneficial insects, ask the company if they were raised at a commercial insectary or wild-caught.

WHAT THEY DO: The adult form is a small white moth that lays a single egg on the bottom of a leaf. The eggs are distinctive: pale yellow and bullet shaped. Once that egg hatches, feeding begins and they make fast work of foliage. They also leave clumps of poop, also called frass, on the plants.

HOW TO GET RID OF THEM: Because these are a perennial problem for me, I'm always on the lookout for the eggs on the leaves of my cabbage family plants; I squish them with a fingernail. I handpick the worms and squish them under my boot. Covers like insect barriers or lightweight row covers are excellent at controlling cabbage worms, as long as they are placed on the plants immediately after planting. With large infestations, the organic insecticide Btk can be applied to affected plants.

Spider Mites

WHO THEY ARE: Spider mites are tiny pests that aren't insects but are (not surprisingly) closely related to spiders. At less than one-twentieth of an inch in length, they're so small, you'll likely need a magnifying glass to spot them. The damage they cause, however, is unmistakable. Their favorite crops include tomatoes, beans, cucumbers, squash, strawberries, and peppers.

WHAT THEY DO: Spider mites suck chlorophyll from the leaves, causing tiny white spots, called stipples, to appear on the leaf surface. Most of the feeding happens on the bottom of the leaves, so often the stippling is the first sign that spider mites have moved into your garden covers. They also produce a fine webbing, another clue to the arrival of spider mites.

HOW TO GET RID OF THEM: Spider mites reproduce quickly and in great numbers, so once you've identified them, don't ignore them. When spotted on houseplants indoors, the general recommendation is to spray the foliage with water as spider mites prefer dry conditions. In a garden

CABBAGE WORMS

SPIDER MITES

structure, wet leaves spread disease so avoid this common advice. If you've struggled with spider mites in the past, avoid placing their favorite plants inside your garden structure. It may be tempting to move some of your indoor tropical plants to the greenhouse for the summer, but you don't want to introduce spider mites to your vegetable plants. To prevent an infestation, I always include flowers and herbs like cilantro and sweet alyssum to attract spider mite predators like ladybugs and lacewings. Predatory mites that feed on spider mites are also available through mail order companies. An organic insecticidal soap spray is another useful tool for reducing populations. Be sure to spray the top and bottom of the leaves and repeat as necessary.

One of the best ways to encourage high pollination rates is to provide plenty of flowers for the bees and other pollinating insects. I include annual flowers like nasturtiums, calendula, sweet alyssum, and zinnias to attract pollinators.

ENCOURAGING POLLINATION

One of the upsides of growing under cover is that it can exclude many insects. Unfortunately, this is also one of its downsides! Crops that rely on insects for pollination may not be sufficiently pollinated when grown under cover, resulting in a lack of fruit. If you're noticing a decreased yield from tomatoes, peppers, cucumbers, melons, and other fruiting crops, you'll want to increase pollinator access to fruiting crops in the future. You can also take steps to prevent poor pollination in the first place.

Open Covers When Crops Bloom

The easiest way to encourage good pollination is to open tops, covers, doors, windows, and roll-up sides of structures on mild days, particularly when vegetables are flowering. Chances are that you'll be opening these up anyway to vent, but if you're using a self-venting garden cover like an insect barrier, be sure to remove that as well when plants that need to be cross-pollinated begin to flower.

Plant Pollinator Habitat

Another easy way to boost pollination is to plant pollinator-friendly flowers and herbs inside and around garden covers to attract pollinating insects. If possible, aim to have flowers in bloom from early spring through late autumn. Your goal should be to create a habitat with a constant supply of food and water to encourage pollinators like bumble bees, mason bees, butterflies, and hoverflies to not just visit, but to call your garden home.

Some of my favorite flowers for the food garden include sunflowers, nasturtiums, zinnias, sweet alyssum, calendula, and cosmos. These can double as cut flowers; or, in the case of nasturtiums and calendula, as edible flowers. Plants in the Apiaceae (carrot) family are very good at attracting beneficial insects. I often grow parsley, dill, and coriander for this reason. If you have space, plant flowering shrubs, perennials, and annuals near garden beds and structures. Just be sure they don't grow tall enough to shade crops.

You can attract more pollinators by placing insect hotels in and around your garden. You can buy them from garden centers or make your own using old logs, hollow twigs and stems, or drilled pieces of untreated lumber. These types of habitats are inviting to native bee species that nest in wood or cavities. It's also a good idea to leave bare patches of earth in and around your vegetable garden and structures where ground-nesting native bees can dig tunnels for their brood cells. These native bees are important pollinators for crops like cucumbers, squash, and pumpkins.

When I spot a just-opened female squash, cucumber, or melon flower, I hand-pollinate by moving pollen from the male flower to the female flower.

Hand-Pollinate When Necessary

Another way to boost fruit production under cover is to hand-pollinate certain crops. Ideally, I'd rather this job go to pollinating insects, but I've learned that to harvest a good crop of melons and edible gourds, I need to give Mother Nature a helping hand.

If I happen to be puttering in my polytunnel and notice open female cucumber, melon, or squash flowers, I'll pop off a male flower and transfer pollen from the stamen to the stigma of the just-opened female flower. You can also use a small, soft paintbrush to transfer pollen from a male flower to a female flower. Touch the dry, clean paintbrush to the male stamen to collect pollen and gently brush it against the stigma of nearby female flowers.

A Refresher Course on
POLLINATION

Pollination is the transfer of pollen from the anther, the male part of a flower, to the stigma, the female part. Pollination leads to the production of fruits, like tomatoes, as well as seeds. There are two main types of pollination: cross-pollination and self-pollination.

Cross-pollination. In order for cross-pollination to occur, pollen from an anther must be transferred onto the stigma of a different flower, but of the same species. Cross-pollination occurs with the help of insects like bees, flies, and butterflies, or with the wind. Cucumbers, corn, and tomatillos all require cross-pollination to produce a harvest.

Self-pollination. Flowers that can pollinate themselves are self-pollinating. Vegetables with self-pollinating flowers include peas, beans, and lettuce.

No pollination needed. Not all vegetables need to be pollinated to yield a harvest, though. For example, broccoli and cauliflower are insect pollinated, but we grow these crops for their immature flower buds. Pollination doesn't need to happen for us to enjoy a bumper crop of broccoli and cauliflower. So, if you use a garden cover to prevent cabbage worms from eating your plants — like an insect barrier floated on mini hoops — you don't need to remove it before harvesting. Also, many of the cucumber varieties I grow are parthenocarpic, which means they are able to produce fruits without pollination. Parthenocarpic varieties are ideal for growing in structures like domes, polytunnels, and greenhouses where pollination may be limited.

PART 2

Vegetables That Love a Cover

For me, the growing season begins in January when the first seed catalogs arrive. But as I begin to plan and dream of the coming year, I'm still harvesting from my mulched beds, mini hoop tunnels, cold frames, and polytunnel. Of course, my garden covers aren't just for winter. They've allowed me to explore a world of vegetables that I never thought possible to grow in Nova Scotia; from burr gherkins to French melons to Thai peppers. Gardening under cover I plant earlier, harvest later, and grow healthier plants that yield a more abundant crop. Here are some of my favorite crops to grow under cover, with tips on how to plant, grow, harvest, and how to best pair them with garden covers.

ARTICHOKE

Oh, how I love artichokes! But they're only perennial in Zones 7 to 10, and I garden in Zone 5. That means I can't grow artichokes, right? Wrong! I've been growing them for more than a decade in my open garden and beneath garden covers. The key is picking a variety like 'Imperial Star', which was bred for annual production, and giving it some strategic protection.

Planting, Growing, and Harvesting

Planting. To ensure you'll enjoy a crop of tender buds in Zones 4 to 6, you'll need to start the seeds indoors. Sow seeds in cell packs and transplant to 4-inch pots when they're 2 to 3 inches tall. I start artichoke seeds under grow lights in late February or early March to ensure good-size artichoke seedlings by planting time.

Growing. Artichokes are normally grown in warmer climates, where they're planted in autumn, overwinter, and start producing buds that first season. To produce buds without going through a cold winter, seedlings will need a chilling period. Expose the plants to temperatures in the 45 to 50°F (7 to 10°C) range for about 10 days. The easiest way to do this is to watch the forecast and when temperatures are reliably above 40°F (5°C), set out the plants. Keep row covers or mini hoop tunnels handy in case frost is in the forecast.

I've found artichoke plants to be prone to aphid infestations, especially when grown in the garden. The good news is that my polytunnel crop is always far less bothered by this common garden pest. If you do spot aphids on your artichokes, a hard jet of water will dislodge most of them or you can spray an insecticidal soap on the plants.

Choose a site with full sun and spend some time on soil prep before you transplant. Artichokes need well-drained soil, plenty of compost, and room to grow; space seedlings 3 feet apart. Ongoing care includes providing a steady supply of water (a soaker hose under a layer of mulch is ideal), pulling weeds, and fertilizing monthly with an organic liquid food like fish emulsion.

Harvesting. Harvest artichoke buds when they've sized up but are still tightly closed. My artichokes are usually 2 to 3 inches long. Cut the stem an inch below the bud. This will encourage smaller, lateral flower buds to form for future harvests.

Cover Strategies

TEMPORARY FROST PROTECTION

Row covers. Use row covers in spring if the weather turns cold while you're giving artichoke seedlings their chilling period.

Cloches. I've had great success using cloches, especially water-filled cloches to protect artichoke plants in spring. Remove plastic or glass cloches during the day for venting and to prevent heat buildup. Replace again at night until the night temperatures are reliably over 50°F (10°C).

Mini hoop tunnels. If you're growing a whole bed of artichokes, mini hoop tunnels are a fast and efficient way to protect the plants in spring. Cover with a row cover or polyethylene. Uncover during the day, re-covering at night if the temperature is forecast to be below 50°F (10°C).

PROTECTED OVERWINTERING

Polytunnels/greenhouses/domes. Plant artichoke seedlings in a polytunnel or greenhouse 3 to 4 weeks before the last expected spring frost. In autumn, further protection may be needed. In my Zone 5B garden I top my polytunnel-grown artichoke plants with 18 to 24 inches of straw covered by a mini hoop tunnel to get them through the winter and remove the protection in early spring. After a few years, the artichoke plants decline, so I remove them.

ARUGULA

Of all the greens I grow — and I grow a LOT — arugula is my number one salad green. My family loves its peppery bite. I grow it practically year-round, taking only a short break in high summer when the heat and dry weather make it more challenging to grow a good crop.

Planting, Growing, and Harvesting

Planting. Arugula goes from seed to harvest in 30 to 40 days. Plant in a sunny or partially shaded site (afternoon shade if possible). Direct sow seed in cold frames or polytunnels 8 to 12 weeks before the last expected spring frost. Space the tiny seeds 1 inch apart in 4- to 6-inch-wide bands. Arugula doesn't hold well in the garden, bolting when the temperature warms, so plan on succession planting fresh seed every 2 to 3 weeks if you want a long season of homegrown arugula. Sow fall and winter arugula in the garden (to be covered with mini hoop tunnels), cold frames, or polytunnels about a month before the first expected frost.

Growing. Germination is super quick, so expect to start thinning just 10 to 14 days after sowing. Our first few harvests are the baby greens from thinning. I aim to thin plants so that they're about 2 inches apart.

Flea beetles are a major pest of arugula. Foil them by rotating crops as far as possible from your last arugula planting and covering newly seeded beds with a lightweight insect barrier.

Harvesting. As soon as leaves are large enough to pick, harvest from the outside of the plant to encourage it to keep growing. Flowers are also edible (and pollinator-friendly), so if your arugula plants bolt and you don't need the space right away, let them bloom.

Cover Strategies

PEST PROTECTION

Insect barriers. Lay over beds just after seeding to prevent flea beetle damage.

DELAY BOLTING

Shade cloths. Shade cloth can delay bolting in spring arugula, extending the harvest by several weeks.

SEASON EXTENSION

Row covers. Lay over plants in spring or fall for frost protection. Row covers can also be floated over the bed on hoops. I use row covers over my winter polytunnel arugula as extra protection.

Mini hoop tunnels. Get an extra-early start on the arugula harvest in spring or extend it into late autumn and winter by planting under a polyethylene-covered mini hoop tunnel.

Cold frames. Low-growing arugula is a great crop for early spring, autumn, or winter cold frames. Vent often in spring and fall to prevent heat buildup.

Polytunnels/greenhouses/domes. Our first arugula harvest comes from our polytunnel, where seed is sown in late February. I continue to plant successive crops until the tunnel begins to heat up in mid-April. By that point, arugula is growing in the garden and doesn't need the polytunnel anyway. In mid-autumn, our arugula once again moves inside the tunnel, where we harvest it all winter long.

'Sylvetta' arugula, also known as "perennial" arugula (even though it's an annual plant), is one of my favorite varieties. It forms 1-foot-tall dense mounds of leaves and often overwinters in my garden. It's slower to grow than garden arugula and has smaller, deeply divided leaves. The flavor is also more intense.

BEANS

Beans are my absolute favorite vegetable to grow and eat (I know, I'm weird!) and using garden covers means I get to enjoy the homegrown bean harvest weeks earlier than my garden-grown crops.

Planting, Growing, and Harvesting

Planting. Add an inch or so of compost before planting, and sow seeds for bush beans 2 inches apart in rows 12 to 18 inches apart. Thin to 6 inches apart. If using a trellis or netting, space seeds 3 inches apart. For pole beans planted at the base of teepees, sow four to five seeds at the bottom of each bamboo pole.

Growing. When bean plants begin to flower and produce pods, provide regular moisture to encourage a high-quality harvest. If plants are water-stressed, it can result in an abbreviated harvest. Ventilate garden structures well to reduce summer heat. If the temperature is above 90°F (32°C), bean flowers may fail to set pods.

My mother always told me to stay out of the bean patch when the weather was wet and she was right! Beans are susceptible to white mold and bacterial blights which spread in wet weather. When irrigating, water the base of the plants if possible, avoiding the foliage.

Harvesting. Once the pods are 4 to 6 inches in length, harvest often — at least every 2 days. If you allow overmature beans to stay on the vines, production will slow. Note that pole beans yield two to three times as much as bush beans when given the same size space.

Cover Strategies

TEMPORARY FROST PROTECTION

Row covers. Drape lightweight row covers over plants if the spring weather turns cool.

Mini hoop tunnels. If cool spring weather turns cold, a polyethylene-covered mini hoop tunnel will keep young bean plants snug and warm. Vent often and remove the cover once the weather has settled.

EARLIER HARVEST

Polytunnels/greenhouses/domes. Plant seed in greenhouses or polytunnels 2 to 3 weeks before the last expected frost. Keep row covers handy and drape them over beds if a hard frost is in the forecast. Compact bush bean varieties like 'Mascotte' offer a relatively quick harvest.

Plastic sheeting. In the open garden, prewarm the soil with plastic sheeting for a week or two before planting, and sow seed after the risk of spring frost has passed.

PEST PROTECTION

Row covers. Lay these on top of newly seeded bean beds to keep birds from digging up the seeds, or drape them over hoops to make a mini hoop tunnel, to protect bean plants from birds, deer, rabbits, and other pests.

SOYBEANS

FAVA BEANS

More Beans for the Polytunnel

Soybeans and broad beans also make excellent crops for polytunnels and greenhouses. Broad beans are tolerant of cool temperatures and should be planted inside your tunnel in early spring. Soybeans, on the other hand, are frost sensitive. Growing them in a polytunnel or greenhouse allows you to plant 2 to 3 weeks before the last spring frost date, or to plant in midsummer for a fall harvest.

BEETS

Earthy and sweet, beets offer a harvest of baby roots and greens in spring and larger, storage-size roots in fall and early winter. They're relatively easy crops and they grow extremely well in early spring or late autumn cold frames, mini hoop tunnels, and polytunnels, offering both an early and late harvest.

Planting, Growing, and Harvesting

Planting. Provide optimum growing conditions so that beets grow quickly. Growth interrupted by dry soil or major temperature fluctuations can result in tough or stringy roots. Ideal conditions include full sun and decent, well-drained soil amended with an inch of compost. Too much organic matter can encourage healthy tops at the expense of good-size roots. The soil should also be slightly alkaline with a pH in the 6.5 to 7 zone. I add kelp meal to my soil before planting, as it contains boron, a mineral needed by beets. Space seed 1 inch apart in rows 12 to 18 inches apart.

'TOUCHSTONE GOLD'

'AVALANCHE'

'CHIOGGIA GUARDSMARK'

'CHIOGGIA'

'GOLDEN'

Growing. Regular irrigation is important for high-quality beets. Thin seedlings when they're about 4 inches tall. Pulling out excess seedlings can dislodge the ones you want to keep, though, so it's better to snip with scissors or pinch them with your fingers. Thin to 3 to 4 inches for baby beets or 5 to 6 inches for larger beets. Use the thinned greens in salads.

Harvesting. Baby beets can be pulled when the roots are just 1 to 2 inches in diameter. I try to pull every second root, leaving the rest to continue growing. Beets won't hold in the soil forever. They become woody and tough if left in the garden too long. Check your seed packet to see how long your specific variety needs to grow from seed to harvest and begin harvesting at that point.

Cover Strategies

EARLIER HARVEST

Cold frames/mini hoop tunnels/polytunnels. Sow seeds 6 to 8 weeks before the last spring frost.

PROTECTION FROM PESTS AND FROST

Insect barriers. Beets are susceptible to leaf miners. Keep the adults from laying eggs on your beet leaves by covering the seeded bed with a lightweight insect barrier.

Row covers. Lay covers on top of the soil at planting time to prevent soil crusting and encourage speedy germination. Also, use them to protect from frost and cold weather in spring and fall.

EXTEND THE HARVEST INTO WINTER

Mulch. Before the ground freezes in late fall, apply a 12- to 18-inch layer of shredded leaves or straw to beet beds in the open garden. Cover with an old row cover to secure the mulch. Harvest into winter.

Cold frames/mini hoop tunnels/polytunnels. Sow a fall crop 6 to 8 weeks before the first expected frost. If your cold frame has a low

profile, pick varieties that don't have tall foliage. Double up your protection and extend the harvest further into winter by mulching cold frame, mini hoop tunnel, or polytunnel beets with a 1-foot-deep layer of shredded leaves or straw in late autumn. Don't forget the foliage! Try growing leaf varieties like 'Bull's Blood' that have cold-tolerant burgundy foliage for winter salads.

BROCCOLI & CAULIFLOWER

Planting broccoli and cauliflower in a polytunnel or greenhouse offers a planting window months earlier than in the open garden, allowing you to foil pests like cabbage worms and enjoy an extra-early harvest. Both crops are cool-season vegetables and grown for their immature flower buds.

Planting, Growing, and Harvesting

Planting. Broccoli and cauliflower produce the highest quality crop when they put on steady growth without drought or heat stress. Meet their ideal growing conditions by giving them rich soil that's well-drained but moisture retentive and full sun. Cool spring or fall temperatures are key to success.

Both broccoli and cauliflower can be direct seeded or transplanted into garden beds or beneath covers. If transplanting, start spring seedlings indoors 4 to 6 weeks before you intend to move them to the garden or garden covers. Because of slug pressure in my garden and occasionally in the polytunnel, I prefer to start with healthy seedlings.

Before sowing seeds or transplanting seedlings, fork in a few inches of compost or rotted

manure. Space seeds 4 inches apart in rows 18 inches apart. Thin to 12 inches once plants are growing well. Space seedlings every 12 inches in rows 18 inches apart. Spacing seedlings 24 inches apart increases head size.

Growing. These crops like cool, consistent conditions, so vent garden structures when the inside temperature hits 50°F (10°C), especially on a sunny day. Steady, even moisture is also important for healthy growth. Stay on top of watering, using soaker hoses and mulch to retain soil moisture. Water-stressed plants can bolt and are more susceptible to insect attacks. Once plants begin to form a tiny head, fertilize with a water-soluble fish emulsion. It doesn't smell so great in an enclosed space like a polytunnel, so apply it when the temperature is mild enough that you can roll up the sides or open the windows and doors. Check frequently for pests like cabbage worms (page 135) and slugs (page 134), handpicking when necessary.

Once cauliflower heads are 2 to 3 inches across, it's time to blanch. Blanching protects the developing head from the sun, preserving the color. Fold the outside leaves over the head and gather them in the middle, securing them with a rubber band. Check often to see how the head is progressing. In spring, cauliflower is ready to harvest about 1 to 2 weeks after blanching. In autumn, it can take 3 to 4 weeks for the heads to reach a harvestable size.

Harvesting. Keep an eye on the broccoli and cauliflower heads as they develop. They go from perfect to overmature in just a few days. Harvest when the heads reach their mature size, typically 6 to 8 inches across for cauliflower and 4 to 6 inches across for broccoli. Harvest size can vary, so refer to your seed packets for specific information about your variety.

Once the main broccoli stem has been cut, if you don't need the space right away, leave the plants to form side shoots. These smaller florets can be harvested for several more weeks. Encourage heavy side shoot development with a dose of liquid organic fertilizer.

Cover Strategies

EARLIER HARVEST IN SPRING

Polytunnels. Start seeds indoors and transplant seedlings into the polytunnel about 6 weeks before the last expected spring frost.

Mini hoop tunnels. Transplant broccoli and cauliflower seedlings 4 to 6 weeks before the last expected spring frost.

PEST PROTECTION

Insect barriers. Broccoli and cauliflower are extremely prone to cabbage worms, which can be controlled with an insect barrier or a lightweight row cover placed over the plants immediately after planting. And I really do mean immediately! I've had the adult cabbage moths flitting around my broccoli transplants *while* I'm planting them. Be sure to leave plenty of excess fabric so that there's enough to accommodate the plants as they grow. Insect barrier and other row covers can also be used to prevent deer and rabbit damage.

LATER HARVEST IN FALL

Mini hoop tunnels/polytunnels. Start fall and winter seeds 8 to 12 weeks before the first expected frost. I start fall and winter crops indoors under grow lights because it's too hot in the polytunnel in late summer to establish healthy seedlings and keep the seedbeds adequately moist.

CABBAGE & CHINESE CABBAGE

Using garden covers, I've found it possible to enjoy homegrown cabbage for around 9 months of the year (although when I consider how much space each plant takes up in the polytunnel and mini hoop tunnels, I'll admit that I do think twice!). But the truth is that I love cabbages of all types — including Chinese cabbage, a cabbage cousin that's also related to turnips — and I'm especially enamored of some of the newly developed cultivars like 'Caraflex', which has a unique pointy head.

Planting, Growing, and Harvesting

Planting. Planting and growing conditions for cabbage are very similar to those for broccoli and cauliflower. Cabbages produce best in a site with full sun with well-drained and well-amended soil. They're also heavy feeders, so I work in some granular organic vegetable fertilizer to the soil at planting time.

It also pays to choose the right varieties for the season. Cabbage and Chinese cabbage are traditionally grown in spring and fall. For a spring crop, choose early or mid-season types so they mature quickly. For autumn cabbage choose a storage-type variety, which usually take longer to mature but yield larger heads.

I prefer setting out transplants over direct seeding (because of the slugs!). If you go this route, start seedlings indoors 4 to 6 weeks before you intend to move them under garden covers. Seedlings should be hardened off and moved to the garden 2 to 3 weeks before the last frost date, or 4 to 6 weeks before the last frost date when moved to garden covers. Set transplants 18 to 24 inches apart, depending on variety and desired

I transplant fall cabbage seedlings in midsummer, when the weather is hot and dry. Using shade cloth helps reduce heat and hold soil moisture so seedlings can get established.

head size. (Refer to the seed packet for variety-specific information.) Rows should be 18 to 24 inches apart.

Alternatively, you can direct seed in garden beds 3 to 4 weeks before the last expected spring frost or 6 to 8 weeks before the frost date beneath covers like mini hoop tunnels and polytunnels. Sow seeds every 4 inches, eventually thinning to 18 inches (miniature varieties of Chinese cabbage can be spaced 12 inches apart).

Growing. The two biggest tasks to remember while growing cabbages under cover are venting and watering. Cabbage and Chinese cabbage are cool-season vegetables that grow best when the temperatures are cool to warm, but not hot. I open the ends of my mini hoop tunnels or roll up the sides of my polytunnel when the inside temperature rises above 50°F (10°C), especially if the sun is out. Good air circulation also decreases the risk of disease.

As with broccoli and cauliflower, use soaker hoses beneath a layer of mulch to maintain even soil moisture. As the crop nears maturity, it's important to be consistent in watering. Prolonged periods of dryness followed by heavy irrigation can cause heads to split. Split cabbages should be harvested right away.

Once the heads begin to form, apply a dose of liquid organic fertilizer. Inspect plants regularly for pests like cabbage worms. In my garden, I've found Chinese cabbage to be a slug magnet, even when they're not an issue on any other crops. Growing them under cover has allowed me to enjoy a harvest of this tasty vegetable without the need to continuously battle slugs.

Harvesting. Harvest cabbage and Chinese cabbage when they reach the size you want, or when they've reached the size indicated on the seed packet. The heads should be firm.

Cover Strategies

EARLIER HARVEST IN SPRING

Mini hoop tunnels. Use a mini hoop tunnel to sow seeds 8 weeks before the last expected spring frost or plant seedlings in the garden 4 weeks before the last expected spring frost.

Polytunnels/greenhouses/domes. If you have the space in your tunnel and a hankering for cabbage, a polytunnel is the best way to start an extra-early spring crop or harvest into late autumn and even winter. Follow the same planting schedule as for growing in a mini hoop tunnel. The exception to this is Chinese cabbage, which prefers the cool, shorter days of autumn over spring. If you do wish to plant Chinese cabbage in your spring structures, don't plant seedlings too early — about 2 weeks before the last expected frost date is ideal. If the temperature dips below 50°F (10°C) for more than a few days, the plants will bolt.

TEMPORARY FROST PROTECTION

Row covers. Row covers are handy for protecting cabbages from frost.

PEST PROTECTION

Insect barriers/row covers. Use these to prevent the moths of cabbage worms and flies of carrot rust maggot from laying eggs on your plants. Remove covers when the weather warms up in late spring.

'MERLOT'

PAK CHOI

RED MUSTARD

Cabbage Cousins: the Asian Greens

There are so many excellent Asian greens to grow in your garden! All of those listed below are cabbage family vegetables and are perfect for spring, autumn, and winter harvesting beneath cold frames, mini hoop tunnels, polytunnels, and greenhouses.

Not only do these Asian greens yield tender, nutritious greens, they also have edible flower buds (like small broccoli florets) and flowers. I often let a few plants bolt to attract bees and supply us with bright yellow blooms for our salads.

PAK CHOI. Pak choi is one of my favorite greens for winter cold frames and our polytunnel. In the open garden pak choi thrives in spring and fall, forming graceful vase-shaped plants with deep green or purple leaves and wide stems that can be white or light green. Use the young leaves for salads, whole or halved baby plants in stir-fries, and mature plants chopped for stir-fries or pickled.

MIZUNA. Mizuna has deeply serrated, almost feathery looking foliage that can be green or green brushed with purple, depending on the variety. It has a muted mustard flavor and we enjoy it in salads, sandwiches, or sliced into pasta and stir-fries. Mizuna is cold tolerant, but it also has more heat tolerance than other Asian greens, with the spring crop usually lasting into summer.

MUSTARD. Beautiful mustard greens come in a wide range of foliage colors, shapes, and textures. They're all cold tolerant with leaves that have a mild peppery flavor when picked as baby greens, but a spicier kick when mature. Plant small, successive crops for a long harvest of high-quality greens.

TOKYO BEKANA. I love this lime green, frilly, non-heading Chinese cabbage. The delicate leaf texture reminds me of leaf lettuce and the mild flavor makes it an excellent addition to salads and sandwiches.

TATSOI. Want winter greens? Grow tatsoi. This low-lying mustard green has a mild flavor and deep green, spoon-shaped leaves that we eat raw in salads or cooked in stir-fries. It's also very fast, with baby plants ready to pull just 3 to 4 weeks from seeding.

CARROTS

Growing carrots under cover is an easy way to make the harvest season last almost year-round. It can give you a serious head start on the planting season in early spring or help raise germination rates of summer-sown seed. Some covers, like insect barriers, help produce a higher-quality harvest by excluding pests like carrot rust flies.

Planting, Growing, and Harvesting

Planting. Carrots grow best in deep, loose, and stone-free soil. I dig in an inch of compost before seeding, but avoid manure, which can cause roots to fork. In the prepared bed, sow seed ¼ inch deep and ½ inch apart. Space rows 6 to 8 inches apart for baby to medium-size carrots. For large storage carrots, space rows 10 inches apart. Pelleted seed results in more even spacing and reduces the need to thin. Keep the seedbed evenly moist until germination, which takes approximately 7 to 21 days.

Growing. Keep the carrot bed free of weeds; fine-textured carrot foliage doesn't compete well.

Once the carrot seedlings are a few inches tall, thin to 1½ to 2 inches apart. Provide consistent, deep moisture to encourage steady growth. If carrot shoulders push out of the soil as they grow, hill soil up around them to prevent greening.

Harvesting. Carrots may be dug any time after they reach the desired size. Peak harvest period lasts about 3 weeks (longer in cool, fall weather), after which the roots may crack or the taste and appearance may decline. Make a few sowings at 3-week intervals for a continuous supply of tender peak-quality carrots. Our favorite carrots of the year are those that are harvested in late autumn and winter after a few hard frosts have converted starches in the roots to sugars, making for exceptionally sweet carrots.

DEEP MULCH

I deep-mulch my carrot beds in late autumn before the ground freezes. A layer of 12 to 18 inches is insulating enough that I can harvest all winter long. To prevent the straw or shredded leaf mulch from blowing away, hold it in place with an old bed sheet or row cover, pinning it down with garden staples.

Cover Strategies

PROVIDING SHADE FOR BETTER GERMINATION

Row covers/shade cloths. These come in handy when carrot seeds are germinating, especially in midsummer when the soil is more susceptible to drying out. Adding a cover on top of the soil reduces evaporation and helps prevent crusting, which is difficult for carrot seedlings to push through. Row covers also capture heat in the spring, both increasing and speeding up germination. If using shade cloth, be sure to remove it once the seeds have germinated.

CROP PROTECTION

Insect barriers. Use insect barriers to prevent damage by carrot rust flies. Carrots don't need to be pollinated to yield a crop, so covers can be left on from when the seeds are sown until harvest.

EARLIER HARVEST IN SPRING

Cold frames. Seed 10 to 12 weeks before the last expected spring frost.

Mini hoop tunnels. Erect a mini hoop tunnel and cover it with polyethylene about 10 to 14 days before you intend to plant. This will prewarm the soil and hasten germination. Seed 6 to 8 weeks before the last expected spring frost.

Best Carrots for Fall and Winter

'NAPOLI'. I've been growing this variety for over a decade in my fall and winter garden as we love the extrasweet roots. They grow 6 to 7 inches long and have smooth roots with a round tip.

'BOLERO'. This has become a family favorite for our cold frames and polytunnel as everyone loves the juicy, crisp roots and sweet flavor. The roots grow up to 8 inches long.

RAINBOW CARROTS. I've overwintered almost every variety of carrot I've grown in a rainbow of colors: purple, red, white, yellow, and orange. While they may not be as sweet as 'Napoli' and 'Bolero', they still taste pretty great, so if you want to deep mulch your bed of 'Atomic Red', 'Purple Haze', or 'Yellowstone' carrots, go ahead.

Greenhouses/polytunnels/domes. Seed 10 to 12 weeks before the last expected spring frost.

LATER HARVEST IN FALL

Cold frames. Seed 10 to 12 weeks before the first expected autumn frost.

Mini hoop tunnels. Seed 10 to 12 weeks before the first expected autumn frost. Erect the tunnel once the temperatures drop close to freezing in late autumn.

Greenhouses/polytunnels/domes. Seed 10 to 12 weeks before the first expected autumn frost.

WINTER CROP PROTECTION

Mulch. In many regions, the winter carrot harvest can be extended by mulching beds in the open garden to keep the soil from freezing. Mulch winter carrots with a 12- to 18-inch layer of shredded leaves or straw before the ground freezes in late autumn. Top the insulating mulch with an old row cover or other piece of fabric to keep the leaves or straw in place. In cold climates (Zones 2 to 4), mulch winter carrots grown in structures.

CELERY & CELERIAC

Celery and celeriac offer similar flavors, but whereas celery is grown for its long crisp stems, celeriac is grown for its large knobby roots. In summer and autumn, we enjoy a crop of celery and in late autumn and winter, our celeriac provides months of aromatic roots.

Planting, Growing, and Harvesting

Planting. Both celery and celeriac, also known as celery root, are slow-growing vegetables that need a head start indoors under grow lights. Sow seed 12 to 14 weeks before the expected planting date. Sprinkle the tiny seeds on the surface of premoistened potting mix and place trays on top of the fridge or on a heat mat. The extra heat can speed up germination. Once seeds sprout, move the trays beneath a grow light.

Hardened-off celery and celeriac seedlings can be planted under a polyethylene-covered mini hoop tunnel a few weeks before the last expected spring frost, or in the open garden once the risk of frost has passed. Keep row covers or polyethylene handy in case the spring temperatures drop below freezing once again.

Cutting Celery

Love the flavor of celery but not the long growing period or fussy demands? Try cutting celery, also known as leaf celery or smallage. It looks like Italian parsley (although the recent variety 'Par-Cel' looks more like curly parsley) but has a strong celery flavor. This is a wilder version of celery and grown for its leaves and flavorful hollow stems. This cool-season vegetable is ideal for autumn garden covers like cold frames, mini hoop tunnels, and polytunnels.

Growing. Celery grows best in rich, water-retentive soil. Before transplanting, dig in several inches of aged manure or compost. Set plants 8 inches apart in rows 18 inches apart. Celery is a water pig, so it's important to provide a steady supply of water as plants grow. Mulching with straw helps soil hold moisture.

I don't bother blanching my celery plants, but blanching does result in lighter colored stems and milder, sweeter stalks. To blanch, hill soil up around plants 2 to 3 weeks before you plan to harvest. You can also blanch stalks by surrounding the plants with a cardboard collar.

Harvesting. Harvest celery stalks all summer long by carefully cutting the outermost stems and leaving the center of the plant to continue growing. Or harvest the entire plant at once, slicing it off just beneath the soil. Celeriac roots can be harvested when they're 3 to 5 inches in diameter.

Cover Strategies

EARLIER HARVEST IN SPRING

Mini hoop tunnels/polytunnels. Celery and celeriac can be grown in a mini hoop tunnel or a polytunnel with seedlings set out anytime from a few weeks before the last expected frost in spring until early summer. The challenge of growing celery under cover is water. It needs consistent water and thrives in moist soil, so mulch your under cover crops.

EXTEND HARVEST INTO WINTER

Mulch. Mulch celeriac roots with a 12- to 18-inch layer of shredded leaves or straw before the ground freezes in late autumn. Top the insulating mulch with an old row cover or other piece of fabric to keep the leaves or straw in place.

CUCUMBERS

Cucumbers are one of the top crops for growing under cover in polytunnels and greenhouses. Greenhouse-grown cucumbers produce higher-quality fruits and offer an extended harvest. They also face less pressure from pests and diseases. Use row covers or mini hoop tunnels to protect garden-grown cucumber seedlings from pests and cold weather.

Planting, Growing, and Harvesting

Planting. If you're growing cucumbers inside a polytunnel or greenhouse, choose a greenhouse variety like 'Corinto', which is vigorous and productive. Cucumbers that are to be grown under cover are usually pruned to control growth. This removes many of the flowers, so choosing gynoecious or parthenocarpic varieties is important to ensure good yield (see The Sex Life of Cucumbers, page 163, for more information).

Before you plant, decide how you want to train and support your cucumber vines, and install any necessary support structures. See the sidebar on page 165 for more information about cucumber support systems.

You can either direct sow or transplant cucumbers. If transplanting, start seeds indoors 2 to 3 weeks before the last expected spring frost. Harden off and transplant to the garden 1 to 2 weeks after the frost date. For protected crops (mini hoop tunnels or unheated walk-in structures), start the seeds indoors 4 weeks before the expected frost date. Cucumbers are cold sensitive, so don't sow or transplant until the soil temperature has warmed up to 65°F (18°C), usually a week or two after the last spring frost. Because they're

The Sex Life of Cucumbers

When flipping through your favorite seed catalog you may notice that the cucumber section is divided into "garden cucumbers" and "greenhouse cucumbers." The difference lies in how many male and female flowers each type has, and whether it needs to be pollinated.

female

male

Garden cucumbers. These have both male and female flowers. For fruiting to occur, insects must move pollen from the male flower to the female flower (the female flower can be identified early on by the tiny fruit beneath its flower). These types of cucumbers are called "monoecious" — usually, there are more male flowers produced than female flowers, which helps ensure adequate pollination of the female flowers. These can be grown in polytunnels and greenhouses, but you'll need to open doors, windows, or roll-up sides during the day to allow bees to pollinate the flowers.

Greenhouse cucumbers. These are either "gynoecious," which produce far more female flowers than male flowers, or "parthenocarpic," which do not need to be pollinated to produce fruits.

Because of the number of female flowers, gynoecious plants have large yields over a relatively short period of time. To help encourage good pollination and enough male flowers, there are usually a few seeds for monoecious varieties included in the gynoecious seed packet. The monoecious seeds are often dyed so you can spot them easily.

Parthenocarpic varieties are the most common "greenhouse varieties." They are generally seedless, but if grown with other varieties and the female flowers are pollinated, the fruits may produce seeds. If you want seedless cucumbers grow only a parthenocarpic variety in a polytunnel or greenhouse, isolating them from other types and screening to prevent pollinators from entering the structure.

so tender, I don't even plant cucumber seedlings in my polytunnel until around the frost date. If the spring temperature dips below freezing in late May, an early planting can easily suffer cold damage and take a long time to rebound. Before planting, enrich the soil with several inches of compost or rotted manure and some organic granular vegetable fertilizer. Follow package directions.

In an open garden, cucumbers that are grown on the ground are typically planted in hills — mounds of soil with four to five seeds planted in each. I grow most of my garden cucumbers vertically on trellises, tunnels, or, in the case of bush types, in cages. Keeping the plants off the ground is an easy way to prevent insect damage, reduce diseases, and encourage good air flow. Space vertically grown plants 12 to 18 inches apart. It's also a good idea to install the trellis before planting. Once seedlings are growing well, it's hard to erect a trellis without damaging the plants.

When direct sowing greenhouse cucumbers, space seed 6 inches apart, thinning to 18 inches. If transplanting, space seedlings 18 inches apart. Plant a second crop of cucumbers 3 to 4 weeks after your first crop. This will take you into late summer and autumn.

Growing. Cucumbers thrive in hot temperatures and quickly put on growth. Keep plants well irrigated so the plants grow unchecked.

For large-fruited cucumbers, remove all flowers and developing fruit from the bottom 2 feet of the plant. This tells the young plant to spend its energy on vegetative growth rather than switching to fruit production. This early sacrifice gives the plants time to size up and you can expect a larger overall harvest. If growing pickling-size or snack-size cucumbers, there's no need to remove fruit from the bottom of the plant.

Pruning and training. Check in with your plants at least once a week to see if suckers need

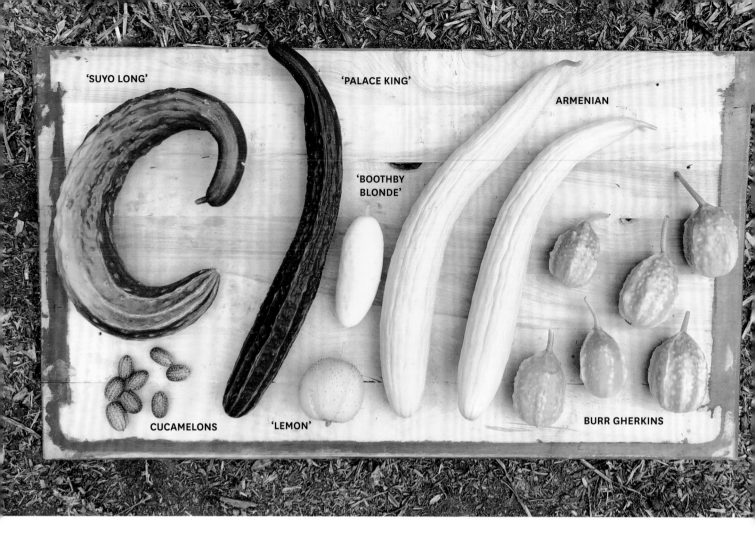

'SUYO LONG'　　'PALACE KING'　　ARMENIAN

'BOOTHBY BLONDE'

CUCAMELONS　　'LEMON'　　BURR GHERKINS

to be pinched or the climbing vines wound around the string. Cucumbers are pruned to produce a more consistent and higher yield. At each node (the spot on the stem where the leaves and suckers emerge) a cucumber plant has one leaf, one tendril, one flower, and one sucker. As the vines grow, allow one main leader to grow, removing all lateral suckers by pinching them off with your fingers.

Once the vine reaches the top of the wire, remove the growing tip and let at least two side buds develop at the top of the plant. These will grow downward, giving you three stems in total producing fruit — the main stem and the two laterals. Continue to pinch off the suckers on these two lateral stems.

Fertilize with a liquid organic vegetable fertilizer every few weeks (check your product label for specific instructions) and vent structures often to promote good air circulation. Keep an eye out for pests and diseases.

Harvesting. Pick fruits often when they reach their ideal size — see variety description for details on mature size. Don't allow overmature fruits to remain on the plants, as this signals a switch from flower and fruit production to seed production. Clip fruits from the vine — don't pull or twist them off.

PINCHING SUCKERS

Cucumber Support Systems

Most cucumbers are produced on vigorous vines, and although you can let them sprawl in the garden, this isn't a good idea in a polytunnel or greenhouse. Not only will they use up a lot of valuable growing space but trellising them gets them off the ground and results in healthier plants.

When growing cucumbers in a walk-in garden cover like a polytunnel, you need a plan for supporting the fast-growing heavy cucumber vines. There are many ways to support cucumber plants, but the two most common methods used when growing under cover are netting or wire and string.

Netting. If you're growing standard garden cucumbers (moneocious varieties) in a polytunnel with no plans to train them to a single stem, you can support the vines with nylon, plastic, or metal mesh netting. Use sturdy metal or wooden stakes to hold the mesh securely or attach it with screws or clips to the wood of one of the gable ends. Choose mesh with holes that are 3 or 4 inches square. Smaller holes can trap developing fruit, causing them to grow into the netting.

Wire and string. Gynoecious or parthenocarpic varieties are usually trained up strings and pruned so that each plant has a single main stem. For this trellising system, you need to run a sturdy wire (I use 12-gauge wire) over the roof trusses, securing it to each gable end. From this wire, tie a length of string or twine above each plant so that the string hangs down to the seedling. Clip the string to the seedling using a tomato clip (available at commercial garden supply stores) or tie it loosely to the base of the plant. As it grows, wind the vine clockwise around the string.

Cover Strategies

EARLIER HARVEST IN SPRING

Plastic mulch and mini hoop tunnels.
Prewarm soil with a clear or black plastic mulch (see page 92) to speed up warming and cover bed with a mini hoop tunnel to allow earlier planting (1 to 2 weeks earlier).

TEMPORARY FROST PROTECTION

Row covers/cloches. If you plant in hills in the open garden, protect newly seeded or transplanted cucumbers with row covers or cloches. Remove cloches during the day to ventilate and prevent heat buildup.

Mini hoop tunnels. A bed of newly planted cucumber seedlings can be protected from sudden cool weather with a quickly erected mini hoop tunnel. Vent well and remove once the weather has warmed up.

PEST PREVENTION

Row covers. Lightweight row covers prevent early damage from pests like cucumber beetles. Remove row covers when plants begin to flower to ensure pollination.

ALTERNARIA LEAF BLIGHT

Cucamelons & Burr Gherkins

Cucamelons and burr gherkins are two of the most popular veggies in our garden — everyone loves their crispy, cucumbery flavors. And since both of these hail from hot climates, using garden covers like row covers or planting them inside a polytunnel in northern climates can mean the difference between no crop and a bumper crop.

Planting. I start both cucamelons and burr gherkins inside under my grow lights about 6 weeks before the last spring frost, a week or two earlier if they'll be planted inside the polytunnel. Sow the seeds in 4-inch pots.

Plant the seedlings a week or two after the last spring frost in a sunny, sheltered spot in the garden. Amend the soil with several inches of composted or rotted manure before planting. Both plants produce vigorous vines and grow best when trellised, tunneled, or given pea and bean netting to climb.

Growing. You may notice no growth for the first few weeks after planting. Don't panic! This is normal as they don't really start to put on growth until the heat of summer settles in. After that, stand back as they'll soon be covering their supports. Provide regular irrigation and mulch plants with straw to prevent weed growth.

Harvesting. Cucamelons are ready to harvest when the fruits are about an inch long. Pick often and don't leave any fruits on the vines to mature until late in the season when you're thinking about seed saving. Older fruits have a sour flavor and will redirect the plants energy from flower and fruit formation to seed production.

Burr gherkins can be picked anytime from chicken egg- to goose egg-size. As with cucamelons, keep on top of the harvest, removing and composting any over-mature fruits. The small spines aren't sharp and you don't need to peel the fruits before eating.

Covers. For plants grown in the open garden, use row covers in late spring if the weather turns cold. Growing inside the polytunnel extends the harvest by about a month in late autumn.

EGGPLANT

Eggplants are heat lovers and a perfect crop for growing under cover in a polytunnel or greenhouse, especially in a region with short seasons. Giving this tender crop a sheltered growing space means an earlier and larger harvest. For those in Zones 3 to 5, you'll increase your eggplant growing success by choosing small-fruited varieties like 'Swallow', 'Patio Baby', and 'Ping Tung'.

Planting, Growing, and Harvesting

Planting. Start seeds indoors 8 to 10 weeks before the last frost date. Bottom heat improves and speeds up germination. Once the seedlings are about 3 weeks old, transplant them to 4-inch pots. A week or two before the last expected frost date, harden off the seedlings and move them into your polytunnel or greenhouse. At this point, the plants are likely to be about 10 inches tall with flower buds forming. Prep your beds before transplanting by digging in some aged manure and a granular organic vegetable fertilizer.

Growing. Space plants 2 feet apart in beds and mulch with straw. If growing in structures, monitor the temperature and vent often for good air flow. The ideal daytime temperature range for eggplants is 80 to 85°F (27 to 29°C) and nighttime range is 62 to 72°F (17 to 22°C).

Eggplants have brittle branches, which can break when they're laden with fruits. Give them support with a tomato cage, stake, or, if growing in a greenhouse, a length of twine running from the crop bars down to the bed or pot. Tie the bottom of the string to the base of the eggplant and wind the string around the stem as it grows. Always wind it in the same direction or else it will unravel and fall down. You can also secure the string with clips (see photo on page 34).

Commercially grown eggplants are pruned to boost yield and encourage air flow through the plant, which can reduce the risk of disease.

I usually let the sucker below the first flower cluster grow so that the plant has two main stems and then clip out all other suckers. Don't be in too much of a rush to remove the suckers; wait until they've produced flower buds. At that point, clip the sucker above the flower buds.

It's also a good idea to remove the foliage below the bottom fruit. These leaves are the first to yellow and can promote the spread of disease. As autumn approaches and the weather begins to chill down, pinch off all the flowers on your eggplants to redirect energy into ripening the existing fruit on the plant.

Harvesting. Harvest eggplants when the fruits are young and glossy. Overripe eggplants have

Choosing a Greenhouse Variety

Although you can grow your favorite varieties inside a structure, you may also want to try a greenhouse variety or two. They're bred to better withstand the wide temperature fluctuations that happen in a polytunnel. They also have a slightly different growth habit, producing taller, less bushy plants, which makes them easier to train.

Another bonus of greenhouse varieties, like 'Angela', is that they have no spines! I'm often skewered by the sharp spikes on the stems and calyx of the fruits when I harvest, and it's a treat to pick eggplant with no fear of being jabbed.

dull skin, developing seeds, and a bitter flavor. Always harvest eggplant with a clean pair of pruners or a knife and a pair of gloves (to avoid getting poked by the spines on the stem and calyx). Slice off the fruit stem so that it's flush with the plant stem.

Cover Strategies

PREHEAT SOIL FOR AN EARLY START AND QUICKER GROWTH

Plastic sheeting and mini hoop tunnels. If you're planting eggplant in the open garden, preheat the soil with plastic sheeting for 2 weeks before transplanting and use a mini hoop tunnel to capture heat for the first few weeks of growth. Open the ends of the tunnel during the day to vent it. Remove the cover once the spring weather has settled and nighttime temperatures are reliably above 62°F (17°C).

TEMPORARY FROST PROTECTION

Cloches. After planting, use cloches to protect individual plants from cold weather. I often use these as nighttime polytunnel protection in late spring when evening temperatures can take a deep dip. Remove or prop open cloches to vent during the day.

PEST PROTECTION

Insect barriers/row covers. If flea beetles are an issue in your garden, cover just-planted eggplant seedlings with an insect barrier or row cover. Place the fabric right over plants or float it on hoops.

EXTEND THE GROWING SEASON

Polytunnels/greenhouses/domes. Using a polytunnel or greenhouse to extend the growing season is the only way many northern or short-season gardeners can produce eggplant.

ENDIVE

Endive is a cool-season salad green and best suited for spring, autumn, and winter salads. The bright, curly leaves of endive look a lot like lettuce, but if you're expecting the mild, sweet flavor of lettuce, you're in for a surprise. Endive is grown for its sharp, bitter leaves that make a tasty salad, especially when paired with cheese, nuts, and a creamy dressing.

Planting, Growing, and Harvesting

Planting. Plant endive seeds or seedlings in a garden bed with full sun and enriched soil. Like lettuce, I direct seed endive early in my garden structures. I seed the first cold frame or polytunnel 10 to 12 weeks before the last expected spring frost and make successive plantings every 2 to 3 weeks.

For a fall crop, I find it easier to start the seeds indoors. They must be sown in mid to late summer, when outdoor conditions are hot and dry, and giving them a few weeks indoors under grow lights promotes healthy seedlings. The germination rate of endive seed plummets when temperatures are above 77°F (25°C). Move seedlings to cold frames, mini hoop tunnels, greenhouses, or polytunnels in late summer to mid-autumn.

There are two main types of endive: frisée and escarole. Frisée forms large, frilly rosettes with narrow, deeply serrated leaves. Escarole has broad, smooth leaves, and is less bitter than frisée.

Growing. Keep the soil consistently moist as plants grow. Thin spring-sown seedlings to 4 to 6 inches apart for baby endive greens or 12 inches apart for full-size heads. Eat the thinnings.

Reduce bitterness by blanching endive rosettes. The easiest ways to do this are by placing an overturned clay flower pot on top of the plant or by gathering up the outer leaves and securing them together with a rubber band. After 1 to 2 weeks, remove the barrier or rubber band and harvest.

Harvesting. Harvest baby greens when the leaves are 3 to 6 inches tall. Harvesting the outermost leaves encourages the rosettes to continue growing and extends the harvest season. For mature heads, harvest by slicing the plant off at soil level.

Cover Strategies

FROST PROTECTION

Row covers. Protect endive from spring or autumn frosts with light or medium-weight row covers. Winter endive grown in greenhouses and polytunnels appreciates an extra layer of row cover in Zones 4 to 6.

DELAY BOLTING

Shade cloths. As with lettuce, shade cloth can delay bolting of spring-grown endive. Float above bed on hoops.

EXTEND THE SEASON

Mini hoop tunnels. Cover endive beds in mid to late autumn with a polyethylene-covered mini hoop tunnel to extend the season by at least 10 to 12 weeks.

Cold frames/polytunnels/greenhouses/domes. Sow seeds in a walk-in structure 10 to 12 weeks before the last expected spring frost and 4 to 6 weeks before the first expected fall frost. Or plant seedlings 4 weeks before the frost date.

FENNEL

Fennel is also called bulb fennel and Florence fennel. The "bulb" isn't a true bulb, however; rather, it's the swollen stem base of the plant. It's grown for its crisp, anise-flavored stems as well as for its ferny foliage.

Planting, Growing, and Harvesting

Planting. Fennel is an easy vegetable to grow, especially when grown as a fall crop; it prefers the shorter, cooler days of autumn. Start with purchased transplants or grow your own by starting seeds indoors late spring through early summer and moving them to the garden after 4 to 6 weeks of growth. Transplant carefully as root damage can initiate bolting. Space seedlings 6 inches apart. I have the greatest success when I transplant seedlings, rather than direct seed (thanks to our local slug population who love the tiny, tender seedlings).

Growing. Fennel prefers full sun and rich, moisture-retentive soil. Dig in some compost or aged manure before planting. Regular irrigation is key to healthy growth; if grown on the dry side, fennel is prone to bolting. Mulch plants with straw to help the soil retain moisture. As the bulbs begin to swell, hill up a bit of soil around them to steady the plants and prevent flopping.

Harvesting. Fennel is harvestable at any size, but I wait until the bulbous stems are at least 2 to 3 inches across. Slice the stem from the root to harvest. If you're lucky, smaller bulbs may regrow from the root stub. You can also let a few plants bolt to collect the aromatic seeds. We use these for an invigorating licorice-flavored tea, but they're also used in cooking and baking. Harvest seed heads once they begin to brown, placing them in a paper bag to try. Once completely dry, shake out the seeds, separating them from any chaff. Store in jars or spice containers.

Cover Strategies

EXTEND THE HARVEST INTO WINTER

Mini hoop tunnels. A mini hoop tunnel topped with row cover is handy for protecting fennel from early to mid-autumn, but switch to a polyethylene cover in mid-autumn to extend the harvest by 6 to 8 weeks.

Polytunnels/greenhouses/or domes. Plant fennel seedlings inside walk-in structures in early to midsummer. Mulch plants with straw and keep them well watered to encourage crisp, high-quality bulbs. Expect to stretch the harvest season by at least 8 weeks with a walk-in structure, even longer if you layer covers by placing a mini hoop tunnel topped with row cover over the greenhouse or polytunnel bed.

KALE & COLLARDS

Kale and collards are four-season vegetables that are easy to grow, productive, and bothered by fewer pests and diseases than most of their cabbage family relatives. I tend to grow more kale than collards and there's not a day of the year we don't have ready-to-pick kale waiting in our garden or beneath garden covers.

Planting, Growing, and Harvesting

Planting. Plant kale based on how and when you'd like to use it. If I want a crop of baby leaves for tender salads, I'll direct sow under cover anytime from late winter to mid-spring, seeding thickly. I repeat this in early and mid-autumn for winter baby greens. Full-size kale or collards can be direct sown, but due to slugs and snails, I prefer to start them indoors, moving the seedlings into the garden 4 to 6 weeks later.

To direct sow kale and collards, plant seeds in cold frames and mini hoop tunnels 6 to 8 weeks before the last spring frost, or 8 to 10 weeks before the last frost in greenhouses and poly-tunnels. For mature plants, space seeds 4 inches apart, thinning to 12 to 18 inches once seedlings are growing well (and eat those thinnings!). For a baby crop, sow seeds densely in bands. I just sprinkle, but if you want to be precise, try to space them 1 to 2 inches apart in a grid pattern.

If you're going to skip spring planting and go right for autumn, start seeds indoors 14 to 16 weeks before the first expected fall frost. Move them to the garden 4 to 6 weeks later, hardening them off first. Or direct sow in garden covers or garden beds (eventually to be covered by mini hoop tunnels) 12 to 14 weeks before the first frost.

Baby kale only takes around a month to go from seed to harvest. For a continual supply of tender baby leaves, succession sow additional bands in your polytunnel or cold frames every 3 to 4 weeks, starting in late summer.

Growing. The key to a high-quality kale or collard crop is consistent moisture; use soaker hoses, water often, and mulch to hold soil moisture.

Harvesting. Kale and collards are ready to harvest whenever the leaves reach the desired size. Harvest individual leaves from the outside of the plant to encourage the center to keep growing and producing. Kale tastes best after a few fall frosts.

I always let a few overwintered kale plants bloom in early spring. Not only does this make the bees very happy, the flower buds (they look like small broccoli florets) and little yellow flowers are edible.

Choosing the Best Varieties

Kale is remarkably hardy with some of the more cold-tolerant varieties surviving temperatures down to –10ºF (–23ºC). Base your kale variety selection on when and how you plan on using your harvest. Certain varieties, like 'Winterbor' (the name gives it away!) are super cold tolerant, while others, like 'Lacinato' (pictured at right), are less hardy. Careful reading of seed catalogs will help you choose the right varieties to grow.

Like kale, collards are cold tolerant (about the same as 'Lacinato'-type kales which are the least winter-hardy kales), but they're also heat tolerant. It's this trait that has made them a classic green in the southern states. For best winter hardiness, grow 'Champion', a collard variety hardy to 15ºF (–9ºC) or 'Even' Star Land Race', a winter superstar that is hardy to 6ºF (–14ºC).

Cover Strategies

PEST PROTECTION

Insect barriers/row covers. Cover newly planted seedlings with insect barriers or row covers to dissuade cabbage worms and cabbage loopers from munching on the leaves of your collard and kale plants in mid to late spring.

TEMPORARY FROST PROTECTION

Row covers. Use lightweight row covers to protect kale and collards from cold weather in early spring or late autumn. Float row covers on hoops. If the fabric is laid directly on foliage, it can freeze to the leaves, causing damage.

EXTEND THE SEASON

Cloches. If you've got a mature kale plant left in your garden by mid-autumn, protect it with a simple DIY cloche. Place a large tomato cage or peony cage on top of the plant and cover with a heavy-duty clear garbage bag. Secure the cover to the metal cage with binder clips or weigh the excess plastic down with rocks or other weights.

Cold frames. Most varieties of kale grow 2 to 4 feet in height and are too tall for my low-profile cold frames. However, they're perfect for temporary straw bale cold frames. If growing kale in lower-profile cold frames like mine, stick to compact 'Dwarf Blue Curled Scotch', which grows just 14 inches tall.

Mini hoop tunnels. Use polyethylene-covered mini hoop tunnels for extra-early kale in March through May and at the other end of the growing season to overwinter mature plants. The 2½- to 3-foot height of the tunnels means they can accommodate most kale or collard plants.

Polytunnels/greenhouses/domes. Because they're so cold tolerant, I generally don't grow mature kale plants inside the polytunnel. I save that space for more tender crops. That said, I do use it for several harvests of baby kale each year, which is one of the first crops seeded in late winter.

LETTUCE

Lettuce is a cool-season salad green that grows best in spring and autumn. It's also a good choice for winter, as there are many cold-hardy varieties that can be grown beneath covers like cold frames or polytunnels. And although lettuce isn't traditionally a summer crop, it can be grown in northern areas if you plant heat-tolerant varieties like 'Coastal Star', 'Jericho', and 'New Red Fire' underneath shade cloth.

Planting, Growing, and Harvesting

Planting. To harvest baby lettuces or by the leaf, sprinkle seeds in a 1-foot-wide band, trying to space them at least an inch apart, and as you harvest the baby greens continue to thin the plants so they are eventually 6 inches apart. I prefer plenty of color and texture, so I plant several varieties in side-to-side bands. For full-size heads, sow seeds every 3 inches in 1-foot-wide bands, thinning to 6 inches apart when the seedlings are a few inches tall. For large-headed varieties like 'Red Sails', thin seedlings to a foot apart. For a nonstop harvest of high-quality lettuce, sow fresh seeds every 2 to 3 weeks from early to mid-spring and again in late summer and early autumn. Our first sowing of lettuce in cold frames and polytunnel beds takes place around the beginning of March, and by the time the weather has warmed up enough for us to seed directly in the open garden, these early plantings are providing daily salads.

Starting seeds indoors gives a jump start on the planting season and provides plenty of seedlings to tuck along the edges of garden beds in containers. Lettuce is not prone to many soil-borne diseases, so it can be planted wherever

I've recently fallen in love with 'Salanova' lettuces, which are both heat and cold tolerant. I harvest them year round from my garden beds, cold frames, and polytunnel.

there is a little open space, without regard to your crop rotation plans.

Growing. Lettuce grows best in full sun, but can also be grown in partial shade, especially during the hot summer months. The best way to grow a fast, high-quality lettuce crop is to provide the plants with ideal growing conditions: consistent moisture, cool temperatures, and a steady supply of nutrients. Add organic matter to the soil before planting and keep soil moist, especially in newly seeded or transplanted beds. Keep an eye out for slugs, snails, and aphids. Handpick slugs and snails and knock aphids from plants with a hard jet of water.

Harvesting. Baby lettuce is ready to harvest when the leaves are about 3 inches long. I pinch the outer leaves by hand, but you can also harvest with scissors, cutting baby plants to about 1-inch-tall stubs. They should grow back several times as "cut and come again" crops. Full-size heads can be picked whenever they reach a harvestable size. Don't wait too long to harvest spring lettuce, as it goes from "Instagram-worthy" to "bolted and bitter" in mere days.

Cover Strategies

EARLIER HARVEST IN SPRING

Mini hoop tunnels. Sow lettuce seeds under a mini hoop tunnel 8 weeks before the last expected spring frost.

Cold frames. Seed 8 to 10 weeks before the last expected spring frost.

Polytunnels/greenhouses/domes. Seed 10 to 12 weeks before the last expected spring frost.

DELAY BOLTING

Shade cloths. Lettuce prefers cool temperatures and plenty of soil moisture and as spring heats up and turns to summer, lettuce bolts. Once lettuce bolts, the leaves become bitter and unpalatable. A sheet of shade cloth draped on hoops over a spring lettuce bed can delay bolting by several weeks.

CROP PROTECTION

Row covers. Lightweight row covers can service lettuce in several ways. First and foremost, they protect from cold temperatures, including light frosts, in spring and autumn. But row covers are also excellent protection from hungry critters like seed-eating birds or lettuce-loving rabbits and deer.

Mini hoop tunnels. Use a mini hoop tunnel topped with polyethylene or row cover to shelter spring lettuce transplants. If hail is in the forecast, a quickly erected mini hoop tunnel offers protection. Come late spring, prolong the harvest and delay bolting by topping hoops with shade cloth.

EXTEND THE HARVEST INTO WINTER

Mini hoop tunnels. A mini hoop tunnel is the perfect cover for autumn harvests. Sow seed or set transplants out around 6 weeks before the first expected frost.

Cold frames. Seed 6 weeks before the first expected autumn frost in cold frames.

Polytunnels/greenhouses/domes. Seed 4 to 6 weeks before the first expected autumn frost in your walk-in structure. For winter harvesting, I add a layer of row cover floated on wire hoops over my polytunnel lettuce beds.

Remember that not all lettuce varieties are equally cold tolerant. For winter harvesting, look to the super cold-tolerant ones like 'North Pole', 'Winter Marvel', 'Winter Density', and 'Arctic King'. Most of these can be harvested all winter, and once the day length extends in late winter, will regrow for early spring salads.

MÂCHE

CLAYTONIA

Great Greens!

If you're looking to take your two- or three-season veggie garden into year-round production, let me introduce you to mâche and claytonia. These cold-hardy greens can be harvested all winter long in Zones 4 and up when grown in cold frames or polytunnels.

Mâche. This has been a staple in our autumn and winter garden structures for many years. The plants form small rosettes — just 2 to 4 inches across — with tiny spoon-shaped leaves. We harvest the rosettes whole by slicing the entire plant off at soil level and enjoy them as a salad green. Seed mâche in mid to late summer in frames and beds. Be mindful that any plants left in the garden by spring soon go to seed and self-sow everywhere.

Claytonia. This is a little-known salad green that deserves its share of the spotlight. Each plant forms a dense clump of small heart-shaped leaves, perfect for fall and winter salads. As the plant matures, the leaves circle the top of the stem and small edible flowers emerge. Sow the tiny seeds directly on the soil surface in early autumn in cold frames and polytunnel beds. Keep beds moist until germination occurs.

MELONS & WATERMELONS

Gardening in a region with short seasons can make it difficult to grow heat-loving melons. But pairing garden covers with early maturing varieties is the best way to enjoy a bumper crop of homegrown muskmelons, cantaloupe, honeydew melons, and watermelons.

Planting, Growing, and Harvesting

Planting. Most types of melons should be started indoors, but Armenian cucumbers, a fast-growing type of muskmelon (not a cucumber!) are the exception to this rule. I both direct sow and indoor sow Armenian cucumber seeds. The transplants go in the garden first and I then direct sow at the same time. This results in a longer harvest window for this family favorite.

For all other melons, start seeds indoors under grow lights about 4 weeks before you intend to plant out (2 to 3 weeks before the last frost date). Bottom heat boosts germination. Once the weather has settled and you've hardened off the seedlings, fork in a few inches of compost or rotted manure in a sunny garden or greenhouse bed. An application of a granular organic fertilizer can also be incorporated into the garden or greenhouse bed at this time.

Don't rush the crop into the ground before the soil has had a chance to warm up; the soil temperature should be around 70°F (21°C). I usually transplant melon seedlings 1 to 2 weeks after the last expected spring frost. Around the frost date, prewarm the garden bed with plastic mulch.

Grow melons vertically to save space and reduce pest and disease problems, especially in polytunnels. Be sure to use sturdy wire or wooden trellises, spacing seedlings 18 inches apart.

Growing. Constant heat is important for healthy growth, so protect just-planted seedlings with row covers or mini hoop tunnels for the first few weeks. Vent covers during the day to allow good air flow. Pull weeds as they appear or apply a plastic mulch (with soaker hose beneath for easy irrigation) to reduce weeds and maintain a warm soil temperature.

Melons and watermelons are monoecious and have separate male and female flowers on each plant. You can give Mother Nature a helping hand by hand-pollinating; just pluck a male flower and gently transfer the pollen to a female flower. Hand-pollinate in the morning if possible when flowers are newly opened, as the quality of the pollen is best at this time. It's also important to open windows, doors, and roll-up sides of your greenhouse or polytunnel to allow pollinators access to the plants.

As they grow, tie the melon vines to the trellis and support the fruits with well-secured slings made of pantyhose, netting, or onion bags. The supports must be breathable and not retain moisture.

Water regularly when the young plants are growing, but try to water in the morning so the foliage isn't wet overnight. Melons can tolerate dry soil, but don't let the plants wilt, which indicates plant stress. Water deeply, but infrequently.

There is big debate about whether melon vines should be pruned. Pruning promotes plant health and results in larger, but fewer fruits. Leaves make sugar and the more leaves on the plant, the sweeter the melons. I'll leave it up to you to decide if you want to prune or not. I pinch out the growth tip once a few fruits have set. And because my season is short, I also pinch off any baby fruits that form after late August. They won't have time to ripen before frost, and leaving them on the vine can make the remaining melons less sweet.

Harvesting. As melons ripen and the harvest date approaches, reduce watering. Irrigating 7 to 10 days before harvest can dilute flavor and reduce sweetness.

It can be hard to tell when a melon is ripe and, though often recommended, thumping it to test for ripeness isn't reliable. Refer to your seed packet for variety-specific recommendations, but one of the biggest signs of ripeness is when the fruits exhibit their mature color. I also lightly press against the fruit where it attaches to the stem. If it falls easily from the stem, it's ripe. Watermelons are usually ripe when the tendril closest to the fruit has withered and dried up and the part of the fruit laying on the ground has turned yellow. Cut — don't pull — the fruits from the vines.

Cover Strategies

TEMPORARY PROTECTION FROM FROST AND PESTS

Row covers. Use row covers to protect young plants from insect damage and cold weather.

Cloches. Individual cloches can be popped over seedlings in spring as cold or bad weather protection. Remove during the day to allow ventilation.

PROVIDE WARMER TEMPERATURES FOR BETTER GROWTH

Mini hoop tunnels. Prewarm garden beds with mini hoop tunnels (or plastic mulch) before planting. Melons love warm temperatures, so leaving a mini hoop tunnel over garden beds for the first few weeks in late spring is extra insurance against bad weather. Vent often.

Polytunnels/greenhouses/domes. A polytunnel or greenhouse provides extra warmth for these heat-loving vegetables, especially when grown in short-season gardens.

If you're growing seedless watermelons, you'll also need to plant a seeded variety for pollination to occur. Most seed packets of seedless watermelons include seeds for the pollinating vine.

PARSNIPS

Growing parsnips is an exercise in patience, but one that pays off come autumn and winter when a harvest of sweet roots awaits. As with carrots, the flavor of parsnips sweetens after several hard frosts in autumn. Parsnips are grown much the same way as carrots, but need a longer growing season, taking almost twice as long to mature.

Planting, Growing, and Harvesting

Planting. A successful parsnip harvest begins with fresh seed; unlike other crops, parsnip seed does not remain viable for years. Prepare the bed the same way as for carrots: loosen soil, remove stones and debris, and dig in an inch of compost. For optimum germination, wait until the soil has warmed to at least 60°F (15°C). Sow seeds an inch apart in rows 18 inches apart. As with carrots, keep the soil consistently moist until germination (14 to 21 days). Once the plants are 4 to 5 inches tall, thin them to 3 inches apart.

Growing. Pull weeds as they appear and hill up soil around the roots if the parsnip shoulders push out of the earth. Maintain a consistent soil moisture. Parsnip foliage contains compounds that can cause a burning rash, particularly on hot sunny days. Those with sensitive skin should be mindful to avoid skin contact with parsnip leaves.

Harvesting. In our house, the parsnip harvest doesn't begin until Christmas when the roots have had weeks of cold temperatures to ensure sweetness. Use a garden fork to *carefully* lift them from the ground; parsnips bruise easily. We harvest as much as we need at one time, leaving the rest to continue to sweeten up. Come spring, dig overwintered parsnips before the tops resprout.

Cover Strategies

EXTENDING THE HARVEST INTO WINTER

Mulch. In many regions, the winter parsnip harvest can be extended by mulching beds in the open garden to keep the soil from freezing. Mulch winter parsnips with a 12- to 18-inch layer of shredded leaves or straw before the ground freezes in late autumn. Top the insulating mulch with an old row cover or other piece of fabric to keep the leaves or straw in place. In cold climates (Zones 2 to 4), mulch winter parsnips grown in structures.

Mini hoop tunnels. Autumn and winter parsnips can be protected with a mini hoop tunnel erected over garden beds before the ground freezes. In Zones 2 to 4, mulch winter parsnips with a 12-to-18-inch layer of shredded leaves or straw before the ground freezes in late autumn, then add a mini hoop tunnel.

Cold frames. Because parsnip seed germination declines in cold temperatures and parsnips won't be harvested until late autumn and winter, I sow cold frame parsnips 3 to 4 weeks after the last spring frost date. In cold climates, a layer of shredded leaves or straw can be added to frames in late autumn.

Polytunnels/greenhouses/domes. I don't often grow parsnips inside my polytunnel as they grow so well in my open garden and I try to save that protected space for high-value crops like tomatoes and peppers. However, if you're a parsnip lover and have polytunnel or greenhouse space, parsnip can be seeded about 120 days before the first expected autumn frost. In Zones 2 to 4, an additional layer of straw mulch is beneficial to winter crops.

CROP PROTECTION

Insect barriers/row covers. Lightweight covers offer frost protection if spring temperatures turn cold, but I mostly use them to prevent damage from carrot rust flies. Parsnips don't need to be pollinated to yield a crop, so leave covers on until harvest, if necessary.

EARLIER HARVEST THE SECOND SPRING

Parsnip lovers impatient for that early spring harvest can speed up the process by erecting mini hoops over the bed in the fall and adding the plastic cover in late winter. This would warm the soil and allow for an earlier harvest.

PEAS

Peas are among the most popular crops grown in my garden beds but they also have a place under cover. My variety of choice is 'Super Sugar Snap', a "mange-tout" pea that offers edible plump pods with small peas tucked within. Super early varieties like 'Patio Pride' are ideal for the spring cold frame or for growing in pots in the spring or fall polytunnel. And if you love pea shoots, be sure to try 'Petite Snap-Greens'. This variety is grown for its edible shoots, flowers, and leaves and is ideal for container growing in walk-in structures.

Planting, Growing, and Harvesting

Planting. Peas are a cool-weather vegetable and should be planted in early spring, about 4 to 6 weeks before the last expected frost. Sow seeds in a sunny garden bed that has been amended with several inches of compost or rotted manure. Plant seeds 2 inches apart in double rows spaced 4 to 6 inches apart. If sowing peas in containers, space the seeds 1 to 2 inches apart in all directions.

Save yourself time and frustration by erecting pea supports before planting. For tall-growing varieties, hang pea and bean netting between well-secured 8-foot-tall stakes. Smaller-growing peas can be held up by a chicken wire trellis or bushy twigs.

Plant peas for autumn by sowing seed 10 to 12 weeks before the first fall frost. Check the "days to maturity" information on the seed packet or in the seed catalog to determine the exact planting date. Be sure to add a week to the "days to maturity" to account for the shrinking days of fall. If a variety needs 60 days from seed to harvest,

Plant trays of fast-growing pea shoots in your polytunnel beginning in very early spring. Scissor-harvest for microgreens when shoots are 4 to 6 inches tall. Sow fresh seed every few weeks for a nonstop harvest.

calculate 67 days just to ensure enough time for the crop to mature.

Growing. Water deeply at least once or twice a week if there has been no rain and pay extra attention to irrigation when the vines begin to flower. Mulch plants with straw or shredded leaves.

Slugs love peas, so I try to stay on top of them by handpicking whenever I see them eating my vines. I also set traps around the plants by mixing water and yeast in small cups set low in the ground. The slugs slither in and drown. I empty the cups every day or two.

Harvesting. Start picking peas when they're a harvestable size and continue to pick every day or two to stay on top of the harvest. Never let overmature pods remain on the vines, as they slow production and reduce the overall harvest.

Cover Strategies

SPEED UP GERMINATION

Row covers. Temporary garden covers like row covers are handy when sowing peas in early spring. The breathable fabric not only keeps birds from eating the just-planted seeds, but also traps heat and moisture to promote quick germination.

EARLIER HARVEST IN SPRING

Cold frames. Get a head start on spring peas by sowing compact dwarf peas like 'Tom Thumb' and 'Patio Pride' in your frames 8 to 10 weeks before the last spring frost. These miniature varieties grow just 4 to 6 inches tall, so they fit well under the low profile of a frame.

Polytunnels/greenhouses/domes. Grow peas in polytunnel beds or pots, planting 2 months before the last spring frost. Sow a round of peas every 2 to 3 weeks from early spring through late spring for the longest harvest. I don't generally grow a lot of peas in my polytunnel, but if I have space in early spring and late summer, I'll sow a crop of 'Super Sugar Snap' to climb netting or the very compact 'Patio Pride' peas in a container.

TEMPORARY PEST AND FROST PROTECTION

Row covers. Use covers to protect seeds planted in the open garden from birds. They can also be used to protect plants from cold weather. Pea seed sown in the garden in mid to late summer for a fall harvest will benefit from a row cover if a cold snap threatens.

Pea shoots grown for microgreens

PEPPERS

Until recently, peppers were one of those vegetables that only did well in my garden during those rare years with long, hot summers. Now, thanks to recently introduced short-season varieties and protective garden structures, peppers are a reliable crop each year.

Planting, Growing, and Harvesting

Planting. Start pepper seeds indoors 8 to 10 weeks before the last frost. Grow lights produce the stockiest seedlings, but you can use a sunny window if it's all you have. Provide bottom heat until germination by placing your pots or seed trays on top of a refrigerator or heating mat.

Harden off and move seedlings to the garden or polytunnel once nighttime temperatures are above 68°F (20°C). Peppers appreciate well-drained, fertile soil, so grow them in raised beds or containers to provide excellent drainage and add compost or rotted manure.

Plant pepper seedlings 18 inches apart and insert a stake or cage at planting time. Pepper plants can be prone to breakage, especially when the branches are heavy with fruits. Providing strong support is an easy way to prevent damage and crop loss. If you're planting in plastic mulch, cut an X in the film for each seedling; run soaker hoses underneath the mulch to make irrigation easy.

Growing. Peppers are slow growing and very sensitive to environmental stress and pressure from insects and diseases. These are plants that like to be pampered, so ensure they have adequate heat, ventilation, fertilizer, and water.

In a greenhouse or polytunnel, peppers can be trained to grow up strings hung from crop bars.

Twist the string around the plants as they grow, always in the same direction. Greenhouse-grown peppers can be left to grow unchecked or supported with a tomato cage, but pruning the plants increases air circulation and promotes a greater yield in terms of both quantity and size. Each pepper node produces a leaf, a flower, and two shoots. The first step is to pinch out the first flower that appears on the plant, called the king flower.

The king flower is the point where the plant begins to branch and two leaders develop. Give each leader a string to support its growth. Every few weeks, pinch out one of the newly sprouted suckers per node, leaving one (remember each node has two suckers). Be careful to not break off the flower buds that form at the nodes.

Harvesting. Bell peppers can be harvested when they reach their mature size and the walls have thickened up — check your seed packet for variety-specific information. They can be picked green or allowed to ripen on the plant for extra sweetness. Ripe bell peppers take an extra 2 to 3 weeks to go from green to their mature color of red, yellow, or orange.

Hot peppers are normally harvested when they reach full size, are firm to the touch, and have developed their ripe color. Jalapeños, however, are picked when they're a glossy, dark green.

Cut peppers from the plant with a sharp, clean pair of pruners or a knife. Cut the fruit's stem flush with the branch.

Cover Strategies

TEMPORARY FROST PROTECTION

Row covers. Keep row covers handy for those spring and autumn days when temperatures slide.

Cloches. After planting, use cloches to protect individual plants from cold weather. Remove or prop open cloches to vent during the day.

PREWARMING THE SOIL

Plastic sheeting. Cover garden beds with black plastic for 2 weeks before planting to prewarm the soil.

PROTECTION FROM PESTS

Insect barriers. Use insect barriers to protect peppers from flea beetles. Bury the edges of the fabric to prevent flea beetles from sneaking beneath the cover.

CREATING A WARM MICROCLIMATE

Mini hoop tunnels. Peppers are heat lovers! Cover seedlings with a polyethylene-covered mini hoop tunnel for a few weeks after planting.

Polytunnels/greenhouses/domes. Hot and sweet peppers thrive in my polytunnel, producing earlier, maturing quicker, and fruiting for a longer period of time. The polytunnel also provides the perfect environment for longer-season varieties that wouldn't have time to mature in my outdoor garden.

RADISHES & TURNIPS

Radishes and turnips are fast-growing vegetables that require the same growing conditions. Traditionally grown for their roots, they also offer edible greens. There are even a few varieties of radishes and turnips grown only for their greens, such as 'Saisai' radish. These are fantastic as a winter cold frame crop. Although spring radishes grow well in spring and autumn, winter radishes like daikon varieties are best planted for fall and winter harvests. They often bolt in early summer before the roots have had a chance to size up.

Planting, Growing, and Harvesting

Planting. Radishes and turnips should be seeded in a sunny spot with well-drained soil. Fork in some compost before planting and loosen soil to a depth of at least a foot if growing large-rooted daikon radishes.

Direct seed radishes and turnips in early spring once the temperature is reliably above 45°F (7°C). I sow in 3- to 4-inch-wide bands, spacing seed 1 inch apart. Neither crop holds for long

in the garden, so only sow a small amount at any one time. Repeat every 2 weeks until late spring.

As plants grow, thin spring radishes by harvesting every second plant when small roots appear. For larger winter radishes like daikon or 'Black Spanish' radishes, thin to 4 inches. Turnips can be thinned in the same manner as spring radishes. For larger turnip roots, thin to 2 inches.

Growing. Hot, dry weather can slow or stunt the growth of radishes and turnips, so plant at the right time and water regularly. If weather heats up, hang a length of shade cloth over the bed.

Harvesting. Most spring radishes are ready to harvest in 3 to 4 weeks. Winter and daikon radishes need twice as long — around 55 to 60 days from seed to harvest. Early maturing turnips like 'Hakurei' are ready in 6 weeks, while later maturing 'Purple Top White' needs 7 to 8 weeks.

Cover Strategies

PROTECTION FROM PESTS AND FROST

Insect barriers. Cover with an insect barrier to deter pesky flea beetles.

Row covers. Use row covers to protect crops from light frost in spring and autumn and to

exclude flea beetles. Float on hoops or lay directly on top of crops.

DELAY BOLTING

Shade cloths. Spring radishes are very sensitive to hot, dry weather and quickly bolt or turn so peppery that it's hard to eat them. Hanging a piece of shade cloth over the bed if the weather heats up delays bolting and can result in a higher-quality crop.

EXTEND THE HARVEST

Mulch. In late autumn, before the ground freezes, turnips and radishes can be mulched to extend the harvest by 6 to 8 weeks. Use a 12- to 18-inch layer of straw or shredded leaves covered with an old row cover.

Mini hoop tunnels. Using mini hoop tunnels in early spring or late autumn extends the harvest by at least 4 to 6 weeks. Vent often to avoid heat buildup.

Cold frames. Spring radishes and turnips make excellent cold frame crops; the first sowing can happen 6 to 8 weeks before the last expected spring frost. Repeat in late summer and early autumn.

Polytunnels/greenhouses/domes. Direct sow 6 to 8 weeks before the last spring frost, sowing more seed every week or two. In late summer to early autumn, start sowing once again.

SCALLIONS

Cold-tolerant scallions are the perfect crop to grow in spring, autumn, and winter cold frames and polytunnels. Also known as bunching onions, these fast-growing, nonbulbing onions have a slender white stem and green leaves. Try mixing seeds of white- and red-stemmed scallions for a colorful crop.

Planting, Growing, and Harvesting

Planting. Hardy scallions can be seeded in late winter in the empty spaces of your cold frame or polytunnel beds. Direct sow seeds 10 to 12 weeks before the last spring frost, spacing them about ½ inch apart. Because they take up such little space, I plant the seed in 4- to 6-inch-wide bands. Sow successive bands every 3 to 4 weeks until late spring. If you want good-size stems, thin young plants to 1 inch apart.

Scallions that are sown in garden beds about 4 to 5 weeks before the spring frost ends can be protected with row covers or mini hoop tunnels if the weather turns cold. In the fall, sow the first

batch of seed in cold frames and polytunnels 8 to 10 weeks before the first fall frost. Seed more in 2 to 3 weeks. For my winter crops, I plant the most cold-tolerant variety, 'Evergreen Hardy White'.

Growing. Young scallions are rather spindly looking plants and don't compete well, so pull any weeds that pop up. Scallions have a shallow root system and regular irrigation is important in producing high-quality plants. If you want the longest possible stalks, hill up soil around the base of the plants several times over the growing season.

Harvesting. Harvesting can begin when the stems are as thick as a pencil. I usually just harvest a few at a time, loosening them with a trowel or hand fork and gently lifting the plants.

Cover Strategies

EXTEND THE SEASON

Mini hoop tunnels. Early spring-sown scallions can benefit from the shelter of a mini hoop tunnel, which will speed up growth and protect from deep temperature dips.

Cold frames. You can grow scallions in cold frames throughout spring, autumn, and winter. Sow a band of scallions in late winter, repeating again in late summer.

Polytunnels/greenhouses/domes. If you have a walk-in garden structure, use it to grow a nonstop harvest of aromatic scallions. Sow seed in small bands continuously from late winter through late summer.

Other Under Cover Onions

Onions grow so well in my garden beds that I don't generally plant them under my garden covers. Plus, most bulbing onions need 3 to 4 months of growing time before they're ready to harvest. That's a lot of precious polytunnel growing space tied up on a vegetable that grows perfectly fine outside.

That said, I do often plant onion family vegetables like scallions, chives, and a few clumps of multiplier onions in my garden structures. They provide fresh onion-flavored greens and, in the case of multiplier onions, small bulbs in autumn. And they help repel insect pests when tucked here and there throughout the tunnel beds. I often cover them with a cloche in early spring to speed up their growth and provide an extra-early harvest. For a large patch, a mini hoop tunnel, erected in early spring, will also do the trick.

Welsh onions. Welsh, or spring onions (*Allium fistulosum*), like other perennial types of onions form a dense clump of hollow leaves. The midsummer flowers are very pollinator friendly. Cut the leaves as needed.

Egyptian onions. Often called walking onions for their ability to move about the garden, this hardy perennial onion (*Allium* x *proliferum*) is easy to grow, even in poor soils. In early spring, the hollow green leaves shoot out of the soil, soon forming a clump about 2 to 3 feet tall. In summer, the stems

are topped with clusters of small ½- to 1-inch bulblets which are deep purple-red. By late summer, the weight of the bulblet cluster causes the stem to bend to the ground, where the small bulblets set roots and form new plants. If not controlled or harvested, this plant will "walk" around your entire garden.

Potato onions. A multiplier type, potato onions (*Allium cepa* var. *aggregatum*) are low-care plants with 2-foot-tall hollow green stems that can be used as scallions. You can also harvest the bulbs from the clump as needed or harvest and divide to plant in a new area.

EGYPTIAN ONION

SPINACH

Spinach is a cold-season vegetable that's ideal for spring, autumn, and winter harvesting. Smooth and arrowhead varieties are perfect for spring and autumn, especially if your goal is to grow tender baby spinach. Savoyed and semisavoyed spinach varieties have crinkly leaves and higher cold tolerance, making them the best varieties for late autumn and winter harvesting.

Planting, Growing, and Harvesting

Planting. Work nitrogen-rich aged manure or compost into the soil before planting. Spinach doesn't transplant well, so direct seed in a 4- to 6-inch-wide band, spacing seeds 1 inch apart. My first seeding in my cold frames and polytunnel is in late February. Subsequent sowings then happen every 3 to 4 weeks until mid-spring. If the soil is workable, sow spinach in the open garden 8 to 10 weeks before the last expected spring frost, covering the bed with a polyethylene-covered mini hoop tunnel.

Late summer, fall, and winter spinach is planted from midsummer to mid-autumn, but sowing spinach in summer when temperatures are hot and the soil is dry results in spotty germination. Boost germination rates by keeping the soil well irrigated and using shade cloth floated on hoops above the bed. Sow spinach 4 to 6 weeks before the first fall frost in garden beds (to be covered with row cover or mini hoop tunnels later in autumn), cold frames, polytunnels, or greenhouses.

Growing. Thin spinach seedlings to 5 inches apart once they're 2 to 3 inches tall and eat the thinnings. Spinach has shallow roots, so water consistently for high-quality leaves. It prefers cool temperatures, so vent structures often in spring and autumn.

Harvesting. Harvest spinach as a baby crop when leaves are 2 to 4 inches tall or anytime they're big enough for you to eat. I try to pick the oldest leaves first so that the plants continue to grow. Keep an eye on your spring spinach and if you start to see signs it is ready to bolt (plants elongating, small flower stalks appearing), harvest and use the whole plant.

Cover Strategies

PROTECTION FROM PESTS

Insect barriers. Use an insect barrier to reduce or prevent damage from leaf miners.

FROST PROTECTION

Row covers. Row covers are handy as frost protection in spring and autumn, and can be used as a double layer of protection inside winter polytunnels and greenhouses.

DELAY BOLTING

Shade cloths. Float a piece of shade cloth on hoops above the bed in late spring to slow bolting or in mid to late summer when seeding an autumn crop.

EXTENDING THE SEASON

Cold frames. Direct sow spinach seed in cold frames 10 to 12 weeks before the last spring frost. Fresh seed can be sown every few weeks for succession harvesting. Start autumn and winter cold frame crops 4 to 6 weeks before the first fall frost.

Mini hoop tunnels. Mini hoop tunnels can be used to shelter spring and autumn spinach, but you can also use it to overwinter fall-seeded spinach for extra-early spring harvests. Sow seed for overwintering about 4 to 5 weeks before the first expected fall frost. Cover with a mini hoop tunnel for winter. Beginning in late February, periodically lift the end of the tunnel to check on the growth. Once the day length is longer than 10 hours, the young plants will grow quickly.

Polytunnels/greenhouses/domes. Sow seed 10 to 12 weeks before the last spring frost, succession planting more seed every few weeks. In autumn, start sowing more seed 4 to 6 weeks before the first fall frost. Spinach can also be seeded just 2 to 3 weeks before the fall frost and overwintered for an extra-early spring crop. In my polytunnel, the spinach I intend to harvest through the winter gets an extra layer of protection in December; I cover the interior bed with row cover floated on wire hoops.

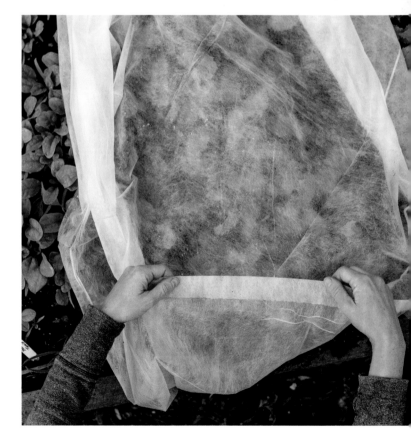

SQUASH

Squash are heat-loving plants that won't tolerate frost or cold weather. For this reason, they're an ideal vegetable to pair with temporary or longer-term covers. They're also prone to insects like squash bugs and squash borers, but putting insect barrier fabrics to work in your garden can reduce or prevent an infestation.

Planting, Growing, and Harvesting

Planting. Squash are greedy plants with high fertilizer needs, so don't skimp on the compost or rotted manure. Dig several inches of it into the bed before planting or prewarming the soil.

Squash seed can be direct seeded or started indoors. Direct seeding should wait until the soil has warmed to 70°F (21°C). Prewarm the soil to encourage quick germination. Planting usually takes place 1 to 2 weeks after the last spring frost.

Sow seeds in the garden in a sunny spot. I plant squash in raised beds, both in the garden and in my polytunnel. Traditional planting advice is to plant in hills that are 1 to 2 feet in diameter,

sowing 3 to 4 seeds per hill (eventually thinning to save just the two strongest plants) or transplanting two seedlings. I don't find this advice practical for small space urban gardens or raised-bed gardens so I don't follow it. My preferred method is to grow winter and summer squash at the back of my garden in straw-manure beds. This is a great use for my half-rotted straw bales that were used during the previous autumn and winter and the squash go crazy for the rich, organic environment. Plus, the mixture of straw and manure holds moisture very well so I rarely water this garden.

If starting squash seeds indoors, use with 4-inch pots, not cell packs which are too small for quick-growing squash plants. Sow two seeds per pot. My planting schedule is to sow indoors 3 to

'ZEPHYR'

'PANTHEON'

'PICCOLO'

Winter Squash in the Polytunnel?

Although I regularly use protective covers like row covers and mini hoop tunnels to establish my squash crop, I don't normally grow winter squash inside my polytunnel. Most varieties take up too much space and they fruit well in the open garden. That said, I've experimented in the polytunnel with some recently introduced compact varieties like 'Butterscotch', a butternut-type squash with short vines and small fruits. I was impressed with the general health of the plants and the yield.

4 weeks before they're hardened off and moved to the garden. Because I'll be using garden covers to protect my squash plants, I set them out 1 to 2 weeks before the last spring frost. Once the seedlings are growing well, snip one off at the soil level, leaving one per pot. (It's hard, but trust me, it pays off!)

If growing vining types up a trellis or netting, direct sow seeds 6 inches apart, thinning to 18 inches. Or you can plant seedlings 18 inches apart, being careful to not disturb the roots when transplanting.

Growing. Squash needs a consistent supply of water for healthy growth, so water often and deeply, and mulch with straw to retain soil moisture. Apply an organic liquid vegetable fertilizer every few weeks to promote strong growth — read packaging for specific application instructions.

If growing winter squash or semivining or vining summer squash like 'Costata Romanesco' or 'Tromboncino' vertically inside a polytunnel, use a sturdy trellis or well-supported netting. Otherwise, opt for bush varieties that stay compact and won't trail into other plants.

Squash are traditionally insect-pollinated but can also be hand-pollinated if necessary. When you notice newly opened female flowers, take a male flower, remove the petals and touch the anther to the stigma of a female flower to transfer pollen. One male flower can pollinate several female flowers.

Harvesting. Male flowers from both summer and winter squash are the first to appear on the vines. These are edible and can be pinched off and eaten. We like to dip them in batter or stuff them with goat cheese and spring herbs like chives and pan fry them.

Once the female flowers appear, summer squashes go from bloom to fruit in mere days! Keep a close watch on the developing fruits and clip them from the vines when they're at peak eating quality. This is 2 to 3 inches for pattypan, 2 to 3 inches for round, and 5 to 8 inches long for zucchini. The exception to this is 'Tromboncino', which should be picked when the fruits are around a foot long.

Winter squashes are long-season vegetables and take months to mature. Don't harvest fruits when immature, as they won't store well and will be susceptible to rot. If there is a frost coming and there are still immature fruits on the vine, harvest them and plan on eating those first.

Remove row cover when squash begins to flower.

Signs of maturity include a hard rind with fruits that sound hollow when gently tapped. The rind should also have turned the mature color for that variety (refer to the seed packet or catalog). Harvest squash before the first frost by cutting the fruits from the vine. Don't twist or pull them from the plants. Leave a 1- to 2-inch stem and handle the fruit carefully. Bruised or damaged fruits don't store well.

After harvesting, cure winter squash in the sun for a week to harden the rind. If a frost is forecast, bring them indoors at night. Store winter squash in a cool basement or garage where the temperature is in the 50 to 60°F (10 to 15°C) range.

Cover Strategies

PREWARM BEDS

Mini hoop tunnels or plastic sheeting. Prewarm garden beds with mini hoop tunnels or plastic sheeting before planting. Squashes love warm temperatures, so leaving a mini hoop tunnel over garden beds for the first few weeks in late spring offers extra insurance if cold weather threatens.

PROTECTION FROM FROST AND PESTS

Row covers. Use insect barriers or lightweight row covers to protect young plants from cold weather and insect damage. If using an insect barrier or row cover to deter pests, cover the bed right after planting. Weigh down the sides to prevent pests from sneaking under the cover. Remove covers when plants begin to flower.

Cloches. Individual cloches can be popped over seedlings in spring as cold or bad weather protection and to deter pests.

EXTEND THE SEASON AND CREATE A WARM MICROCLIMATE

Polytunnels/greenhouses/domes. Growing summer squash inside a polytunnel is a great way to enjoy an earlier harvest; you'll get a jump on the season by 2 or 3 weeks. I grow bush types in my polytunnel beds, spacing them about 24 inches apart. I plant two successive crops for a midsummer harvest and a mid-autumn harvest. If your space is limited, try planting a compact variety of winter squash. 'Honey Bear' and 'Butterscotch' are excellent small, disease-resistant varieties.

SWISS CHARD

Swiss chard is a green that keeps on giving, producing nonstop from mid-spring through late autumn — longer if grown under cover! And with its colorful foliage and stems, it's the poster crop for ornamental edibles. I adore how 'Peppermint', 'Orange Chiffon', 'Rhubarb', and 'Bright Lights' chard add color to the garden and polytunnel.

Planting, Growing, and Harvesting

Planting. Swiss chard grows well in full sun to part shade, preferring some shading from the hot summer sun. It also likes a rich soil, so dig in several inches of aged manure or compost before planting.

Sow seeds 2 inches apart and ½-inch deep in rows 18 inches apart. Swiss chard "seeds" are actually fruits with a cluster of several seeds. Thin to 6 inches for baby chard or 12 inches for full-size plants. Seeds can be started indoors under grow lights 4 to 6 weeks before you plan on moving them to the garden or garden covers.

Growing. Swiss chard is a trouble-free crop, with leaf miners perhaps the biggest complaint. Make sure to water deeply at least once or twice a week. Regular moisture results in healthy plants and more tender leaves. Mulch soil to hold moisture.

Harvesting. Harvest by picking leaves from the outside of the plant. For baby greens, pick leaves when they're 3 to 5 inches long. If older leaves yellow, remove and compost them to encourage fresh growth.

Cover Strategies

PROTECT FROM PESTS

Insect barriers. An insect barrier helps prevent adult leaf miners from laying eggs on the leaves. Be sure to weigh the cover down or bury the bottom edge under the soil. If there are any gaps, adult flies can sneak in.

PROVIDE SHADE FOR A HIGH-QUALITY HARVEST

Shade cloths. Maintain optimum quality by placing shade cloth over your chard patch in midsummer.

EXTEND THE SEASON

Cold frames/mini hoop tunnels/polytunnels. Swiss chard can be direct sown in early spring in cold frames, mini hoop tunnels, and polytunnels 8 to 10 weeks before the last frost. Grow chard as a baby crop in low-profile cold frames. Mature chard plants are too tall for my wooden cold frames, but they can be protected with a straw bale cold frame in mid-autumn or grown in polytunnels and mini hoop tunnels.

TOMATOES

Growing tomatoes under cover in a walk-in structure offers a lot of advantages: earlier and later harvesting, reduced pressure from soilborne diseases, and a heavier crop. Even just pairing tomatoes grown in open garden beds with simple season extenders will help you get a good start on the growing season and protect plants from frost.

Planting, Growing, and Harvesting

Planting. Sow seeds under grow lights 6 to 8 weeks before you intend to move the plants to the garden or greenhouse. Prep the garden or polytunnel bed for planting by amending the soil with several inches of compost or aged manure and a granular organic tomato fertilizer (follow package application directions).

Plant fully hardened-off tomato seedlings deeply — up to the first set of leaves. Unlike most vegetables, tomatoes will root all along their stem, producing a more robust plant.

Spacing depends on variety. Plants that are to be trained vertically can be planted closer than those allowed to sprawl. Most of my tomatoes are planted around 2 feet apart. It's better to space them farther apart than to overcrowd them, which can lead to poor air circulation and disease. Mulch newly planted tomato with straw or shredded leaves.

Growing. Growing a bumper crop of mouthwatering tomatoes requires full sunshine, even moisture, well-amended soil, proper support, and good air flow. Vent structures often, whether you're using mini hoop tunnels in spring or polytunnels in summer. If the temperature climbs past 90°F (32°C), pollination declines and blossom drop can occur. No pollination means no tomatoes. Venting a polytunnel or greenhouse also allows bees and other insects to pollinate your tomatoes.

There are many ways to support greenhouse- and garden-grown tomatoes. They can be caged (if the cages are tall and extrasturdy), staked, trellised in a basket weave, or supported by twine or strings. In a polytunnel or greenhouse, the last technique is preferred, especially if you have crop bars, which allow a heavy-gauge metal wire to be run between the gables. The string is tied to the metal wire and tied or clipped to the base of each tomato plant. As the plant grows, the string is wound around the stem or clipped with a tomato clip (available at greenhouse supply stores). See page 34 for more information.

Polytunnel plants that are grown vertically and relatively tightly need to be pruned to improve air flow, reduce the risk of disease, and maximize production. Determinate tomatoes need little pruning, but I do pinch out the suckers below the first flower cluster to encourage a strong central stem. Indeterminate tomatoes are usually pruned according to their supports: caged tomatoes are lightly pruned, staked plants are moderately pruned, and trellised or string-supported tomatoes are heavily pruned.

I prune my string-supported polytunnel tomatoes to a single stem by pinching out all the suckers that form. If you stay on top of it, removing young suckers every 10 to 14 days, it's a quick task to do by hand. If you skip a few weeks, you'll find clean, sharp pruners make quick work of pruning.

The last bit of pruning is topping the plants in late summer to remove the flowers and immature fruits that won't have time to mature before the end of the growing season. This speeds up ripening for the remaining tomatoes on the plant.

PRUNING A SUCKER

Growing Under Cover Tomatoes in Containers

There are many container-friendly tomato varieties that are excellent tucked into pots or hanging baskets wherever you have a bit of space in the polytunnel or greenhouse. My favorites include 'Terenzo', 'Sweetheart of the Patio', and 'Tumbler'. Container-grown tomatoes in polytunnels and greenhouses dry out quicker than those grown in beds, though, so water often and never let plants get to the point of wilting.

Harvesting. Harvest fruits as they ripen. It can be tricky to tell when unusually colored varieties, like 'Cherokee Purple' or 'Great White Blues' are ripe, but I recommend going by color and feel. The fruit should be the mature color described or pictured on the seed packet and it should be firm but have a bit of give when gently squeezed. Harvest ripe tomatoes by clipping them from the vine. Cherry tomatoes can be gently pulled off plants.

'COSTOLUTO GENOVESE'

'GARDEN PEACH'

'MOUNTAIN MAGIC'

'JAPANESE BLACK TRIFELE'

Choosing the Right Varieties

Your homegrown tomato harvest starts with picking the right varieties to grow. If you're going to be growing tomatoes in a polytunnel, greenhouse, or dome, you may want to try to experiment with greenhouse varieties. These are generally tall, indeterminate plants (to produce a larger harvest) and bred to be grown in the protective environment of a greenhouse. They also offer improved disease resistance.

Of course, you don't want your homegrown tomatoes to taste like supermarket tomatoes (bred for shipping and storage, NOT flavor), so read seed catalogs carefully. Good-flavored greenhouse varieties include 'Estiva', which has medium-size red fruits; 'Apero', a cherry-grape cross with sweet red fruits; and 'Clementine', a cocktail tomato with golden orange fruits. Expect the seed to cost more than garden varieties.

Cover Strategies

PREWARM THE SOIL

Plastic sheeting. In outdoor beds, prewarm the soil with black plastic film while the seedlings harden off.

PROTECT FROM INSECTS AND FROST

Row covers. As with other members of the tomato family, a row cover or insect barrier helps shield the plants from frost, cold weather, and insect damage.

Cloches. If you only have a few tomato seedlings, cloches are handy in spring for protecting them when the temperature drops. In autumn, mature determinate tomato plants can be protected with a large DIY cloche — a tomato cage covered with a heavy-duty garbage bag when day or night temperatures go below 50°F (10°C).

Mini hoop tunnels. These are my go-to protection for spring tomato seedlings. A PVC or metal mini tunnel goes up quickly and can be topped with row cover or polyethylene to shield the cold-sensitive plants. I generally place a cover over newly planted tomato seedlings for 2 to 3 weeks, venting the ends during the day.

EXTEND THE SEASON

Polytunnels/greenhouses/domes. A game changer for tomato lovers gardening in short-season regions! Walk-in structures allow earlier planting, a protected growing environment, earlier and extended cropping, and fewer pests and diseases.

Tomato-Family Relatives

Potatoes. These aren't traditionally greenhouse or polytunnel crops, but you can tuck a few seed potatoes in beds or containers in a walk-in structure 4 to 6 weeks before the last spring frost. Choose early maturing varieties and aim to harvest them as baby potatoes.

In the open garden, row covers or insect barriers can be used to protect potato plants from frost or cold weather, or from pests like the Colorado potato beetle. To be effective, rotate potatoes as far as possible from the previous year's crop and cover the bed with the fabric right after planting. Weigh down or bury the edges of the fabric, leaving slack for the plants to grow.

Ground cherries and tomatillos. These are two tomato relatives you might want to try growing under cover! Both plants resemble tomato plants, but produce their fruits in papery husks — ground cherries are marble-size and relatively sweet, while tomatillos are closer to golf ball size and have a tart, citrusy flavor.

Both ground cherries and tomatillos can be afflicted by flea beetles, so insect barriers and row covers can be useful when growing them in the garden. You can also use mini hoop tunnels to protect them from frost and inclement weather.

ACKNOWLEDGMENTS

I'm so grateful for the support of my family: Dany, Alex, and Isabelle, as well as my mother, Joyce, and in-laws, Kamal and Noha. Also Lisa, Jason, Ryan, and Lucy; Jean-Louise, Nick, Brayden, and Jackson; and Tony, Leah, Sophia, and Mya.

Thank you to all the wonderful folks at Storey Publishing. I count myself very lucky to be a part of the Storey family, an inspiring and talented group of people who strive to make the world a better place. Thank you to Carleen Madigan, the brilliant editor of this book, as well as of *Veggie Garden Remix* and *The Year-Round Vegetable Gardener*. I'm grateful for the creative vision of art director Carolyn Eckert and production designer Erin Dawson. And thank you to Colleen Mulhern, the associate marketing and publicity manager for *Growing Under Cover*.

Massive thanks to my good friends and SavvyGardening.com partners Jessica Walliser and Tara Nolan, who inspire me every single day.

It was a treat to work with Jeff Cooke and Jenn Nauss of Cooked Photography. It's no easy task to make a row cover or sheet of polyethylene look glamorous, but they stepped up to the challenge and made every photo shoot a garden party.

Thank you to all the home gardeners who tag me in social media posts or send me photos of their DIY greenhouses, cold frames, garden domes, and various garden structures. A few of these you'll find in the book, including Stephen Farley's pallet greenhouse and Steve and Jeani Mustain's wood-framed greenhouse. And I'm always inspired by the beautiful food garden of Rob and Brenda Franklin, which uses covers to stretch the season into a year-round harvest.

I'd also like to thank and acknowledge some of the companies that provided products or photos for the book.

The entire Gardener's Supply Company catalog is basically my wish list. In this book you'll find their three-season plant protection tent, super hoops, pop-up tomato accelerator, all-purpose garden fabric, and willow cloches.

The folks at Johnny's Selected Seeds are wonderful to work with and have supplied me with row covers, polyethylene sheeting, and a four-foot low tunnel hoop bender.

The Halifax Seed Company has been a source of many of my garden covers over the years. I've purchased Lexan polycarbonate for cold frames, various weights of row covers, plastic soil mulch, and my 14-by-24-foot polytunnel from them.

The staff at my local Lee Valley Tools store, as well as at their head office, have been extremely supportive of my work. The polycarbonate cold frame in my garden is their double-walled cold frame — a handy, versatile garden cover.

I've been a big fan of Smart Pot fabric planters for years; I use their six-foot- and eight-foot-long big bag raised beds in my garden and polytunnel.

Hartley Botanic is the manufacturer of my dream greenhouses, and I am pleased they shared several photos of their structures in this book.

METRIC CONVERSIONS

TO CONVERT	TO	MULTIPLY
inches	millimeters	inches by 25.4
inches	centimeters	inches by 2.54
inches	meters	inches by 0.0254
feet	meters	feet by 0.3048
feet	kilometers	feet by 0.0003048
yards	centimeters	yards by 91.44
yards	meters	yards by 0.9144

INDEX

Page numbers in *italic* indicate photos or illustrations; numbers in **bold** indicate charts.

system set-up, 109. *See also* heat retention systems; irrigation; ventilation

  heat, providing additional, 116–17, *116*, *117*

  humidity, 110–11

  temperature, 110–11

T

tatsoi (mustard greens), 155

temperature, controlling, 110–11, *110*

temporary structures, 20

thermal mass, structure cooling and, 120

Tokyo Bekana (Chinese cabbage), 155

tomatillos, 207, *207*

tomato blight, early, 133

tomatoes, 202–7

  'Apero', 205

  'Bellini', 133

  cages for, 88

  'Cherokee Purple', 204

  'Clementine', 205

  in containers, under cover, 204, *204*

  'Costoluto Genovese', *205*

  cover strategies for, 206, *206*

  'Defiant', 133

  'Estiva', 205

  'Garden Peach', *205*

  'Geronimo', 133

  'Great White Blues', 204

  growing, 203, *203*

  harvesting, 204

  'Japanese Black Trifele', *205*

  'Jasper', 133

  'Lizzano', 88

  mini hoop tunnel over, 83, *83*

  'Mountain Magic', 133, *205*

  'Mountain Merit', 133

  in nightshade family, 105, *105*

  planting, 203

  pruning suckers, 204, *204*

  red plastic mulch for, 92, *92*

  stakes for, 88, *88*

  'Sungold', 82, 88

  support systems for, 86, *86*, 87, *116*

'Sweetheart of the Patio', 204

'Terenzo', 204

tomato family relatives, 207, *207*

'Tumbler', 204

varieties, selecting right, 205, *205*

trellises, 88

tripods, 88

tunnels. *See also* mini hoop tunnels; polytunnels

  mini tunnels, 19, *19*

turnips, 190–91

  'Hakurei', 190

  'Purple Top White', 190

U

underground greenhouse (walipini), 74–75, *74*, *75*

V

vacation, planning for, 127

vegetable families

  crop rotation and, 103–4

  guide to, 105, *105*

vegetables, 140. *See also* crops; *specific type*

  heavy vining, 54

  ideal air temperatures for, 110

  soilborne diseases and, 104

ventilation, 118–120

  cold frames, 43, 44, *44*

  disease prevention and, 132

  manual versus automatic, 120, *120*

  mini hoop tunnels, 34–35, *35*

  other cooling methods and, 120

  overwintered crops and, 100–101

  ways to vent, 118–120

vertical containers, 89

vertical growing, 86–89, *86*, *87*, *88*

  benefits of, 86–87

  best crops and systems for, **88**

  crop bars for, 87

  fruits, supporting larger, 89, *89*

  methods, additional, 88–89

  planning for, 87

  pruning plants and, 89

  shade and, 87

supportive structures, additional, 88

training plants, 87–88

twine for, 87, *87*

wire-and-string method, 87–88

W

walipinis, 74–75, *74*, *75*

watering. *See also* irrigation

  cold frames, 43–44, *43*

  disease prevention and, 132

  fertilizers and, 126

  garden paths, cooling and, 120

  by hand, small structures and, *121*, 123

  mini hoop tunnels and, 35

  overwatering and, 122

  rain barrels and, 121, *121*

  rainwater collection, 71

  site preparation and, 52

  soaker hose and drip tape, 124, *124*

watermelons, 182–83, *182*

  cover strategies for, 183

  growing, 183

  harvesting, 183

  planting, 182

water reservoirs, 112

weather

  extremes, protection from, 17, *17*

  watering with the, 122

  windbreaks and, 114

Welsh onions, 193

whiteflies, *134*, 135

windbreaks, 114

winter. *See also* snow

  mulching in, 17, *17*

  planting crops for, 93

  polytunnels in, 52, *52*, 53

winter crops

  caring for overwintered, 100–101, *100*, *101*

  inside polytunnels, 67, *67*

wire tomato cages, 88

wooden tomato cages, 88

GATHER GREAT GARDENING ADVICE

with More Books by Niki Jabbour

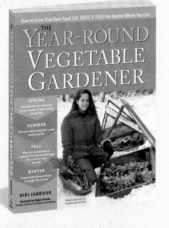

The Year-Round Vegetable Gardener

Grow your own food 365 days a year, no matter where you live! Learn the best varieties for each season, master succession planting, and make inexpensive protective structures that keep vegetables viable and delicious through the colder months.

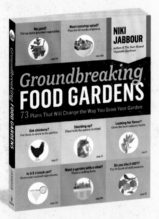

Groundbreaking Food Gardens

This stellar collection of 73 surprising garden plans from leading gardeners highlights unique themes, innovative layouts, unusual plant combinations, and space-saving ideas. These illustrated designs are sure to ignite your creativity!

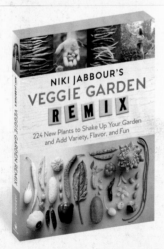

Niki Jabbour's Veggie Garden Remix

A lively "Like this? Then try this!" approach starts with what you know, then helps you expand your vegetable gardening repertoire by suggesting related varieties and offering detailed growing information for more than 200 plants from around the globe.